# McGraw-Hill
# Illustrative
# Mathematics®

## Algebra 1 Supports

McGraw Hill

**Cover credit:** piranka/Getty Images

# mheducation.com/prek-12

Send all inquiries to:
McGraw-Hill Education
STEM Learning Solutions Center
8787 Orion Place
Columbus, OH 43240

ISBN: 978-1-26-421067-1
MHID: 1-26-421067-1

*Illustrative Mathematics, Algebra 1 Supports*
*Student Edition, Volume 1*

Printed in the United States of America.

1 2 3 4 5 6 7 8 9 10 11 12 LMN 28 27 26 25 24 23 22 21 20

'Notice and Wonder' and 'I Notice/I Wonder' are trademarks of the National Council of Teachers of Mathematics, reflecting approaches developed by the Math Forum (http://www.nctm.org/mathforum/), and used here with permission.

# Contents in Brief

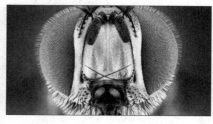

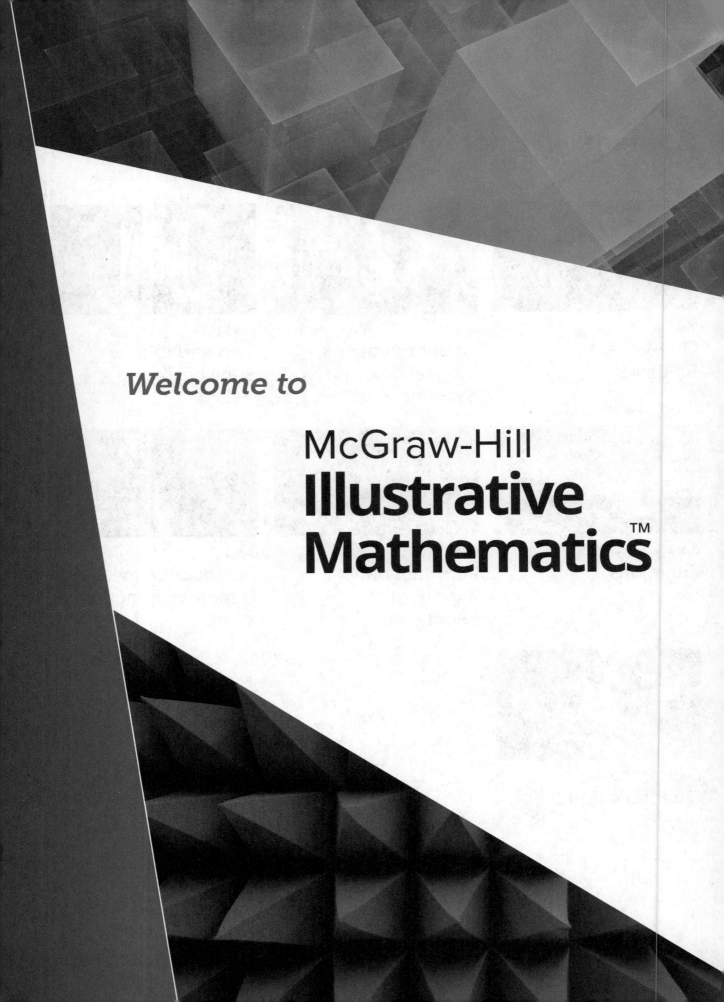

*Welcome to*

# McGraw-Hill
# Illustrative
# Mathematics™

*Illustrative Mathematics* is designed to provide opportunities for you to learn math by doing math. By engaging in problem-solving activities, you will make and test your own conclusions and try out different strategies as you learn math.

## Let's get started!

**Unit 1**

# One-variable Statistics

Kingarion/Shutterstock

# Linear Equations, Inequalities, and Systems

## Unit 3

# Two-variable Statistics

## Unit 4
# Functions

**Unit 5**

# Introduction to Exponential Functions

Unit 6

# Introduction to Quadratic Functions

© IongKo Image Stock/Alamy Stock Photo

**Unit 7**

# Quadratic Equations

# One-variable Statistics

Gardeners can use data gathered from the soil to determine how much fertilizer needs to be applied. You will learn more about analyzing data in this unit.

## Topics
- Getting to Know You
- Distribution Shapes
- Manipulating Data
- Analyzing Data

# One-variable Statistics

Lesson 1-1

# Human Box Plot

NAME _____ DATE _____ PERIOD _____

**Learning Goal** Let's recall how to create and interpret a box plot

 ## Warm Up
### 1.1 Math Talk: Subtraction

Evaluate mentally.

$100 - 20$

$100 - 25$

$100 - 73$

$100 - 38$

 ## Activity
### 1.2 Human Box Plot

Using the list of data, find the lower quartile, median, upper quartile, lowest value, and highest value of the data set. Follow your teacher's directions to create a human box plot.

# Activity

## 1.3 Creating Box Plots

With a partner, calculate the five-number summary and create a box plot of your data set. When you are done, include the name of your data set and display your box plot for the class to see. Then, find the answers to the questions below from your classmates' box plots.

1. Which data set's greatest value is 10?

2. Which data set has the largest range?

3. Which data set has a median of 16.5?

4. Which data set has the greatest upper quartile value?

5. Which data set's least value is 3?

6. Which data set has the smallest range?

Lesson 1-2

# Human Dot plot

NAME _____ DATE _____ PERIOD _____

**Learning Goal** Let's recall how to create dot plots.

## Warm Up

**2.1 Notice and Wonder: Flipping Coins**

200 students flip a coin 100 times and record the number of heads that are flipped.

What do you notice? What do you wonder?

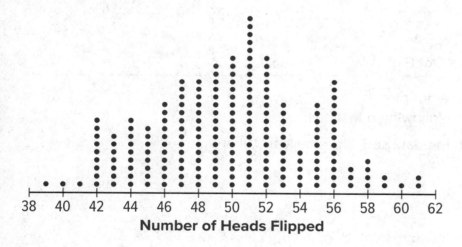

| Number of Heads Flipped | Frequency | Number of Heads Flipped | Frequency | Number of Heads Flipped | Frequency |
|---|---|---|---|---|---|
| 39 | 1 | 47 | 14 | 55 | 11 |
| 40 | 1 | 48 | 14 | 56 | 14 |
| 41 | 1 | 49 | 16 | 57 | 3 |
| 42 | 9 | 50 | 17 | 58 | 4 |
| 43 | 7 | 51 | 22 | 59 | 2 |
| 44 | 9 | 52 | 17 | 60 | 1 |
| 45 | 8 | 53 | 11 | 61 | 2 |
| 46 | 11 | 54 | 5 | | |

## Activity
### 2.2 Human Dot Plot

1. Follow your teacher's directions to create a human dot plot.

2. Create a dot plot that represents the same data as the human dot plot.

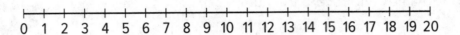

0  1  2  3  4  5  6  7  8  9  10  11  12  13  14  15  16  17  18  19  20

## Activity
### 2.3 Constructing a Dot Plot

Using the class data, construct a dot plot.
Use your dot plot to answer the following questions:

1. What is the largest value in the data set? The smallest? What do these numbers represent?

2. What is a typical amount of sleep for a student in your class?

3. It is recommended that teenagers get 8–10 hours of sleep each night to perform at their best the following day. Based on the data, how well do you think your class would perform on a test? Explain your reasoning.

4. What would the dot plot look like for a class that has the same number of students, but those students tend to get less sleep than students in your class?

Lesson 1-3

# Which One?

NAME _____ DATE _____ PERIOD _____

**Learning Goal** Let's explore the best uses of box plots, dot plots and histograms.

 ## Warm Up

### 3.1 Math Talk: What Was the Final Temperature?

Mentally evaluate the final temperature in each scenario.

The temperature was 20 degrees Celsius and it dropped 18 degrees.

The temperature was 20 degrees Celsius and it dropped 20 degrees.

The temperature was 20 degrees Celsius and it dropped 25 degrees.

The temperature was 20 degrees Celsius and it dropped 33 degrees.

Use the class data to create a dot plot, box plot, and three histograms, each with different bin sizes.

1. Create a dot plot.

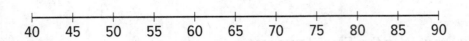

2. Create a box plot.

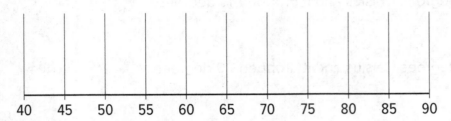

3. Create a histogram using intervals of length 20.

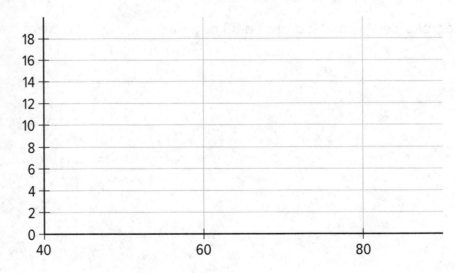

NAME _____ DATE _____ PERIOD _____

**4.** Create a histogram using intervals of length 10.

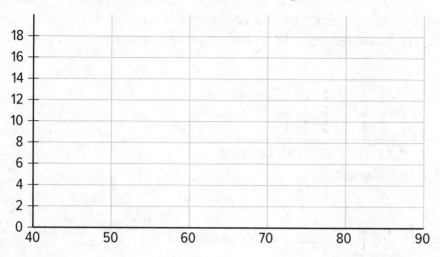

**5.** Create a histogram using intervals of length 5.

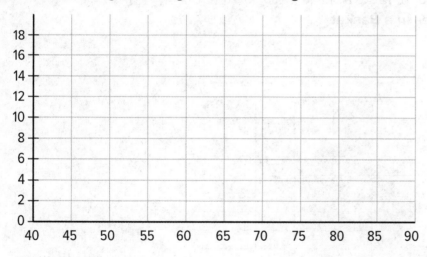

**6.** Which of these representations would you use to summarize your class' data: the dot plot, the box plot, or one of the histograms? Explain your reasoning.

### 3.3 Which One?

There are several baskets on a table, and each basket contains a certain number of strawberries. Here are three data displays showing the number of strawberries in each basket.

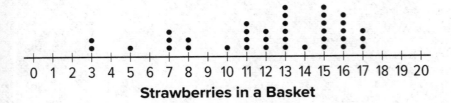

**Strawberries in a Basket**

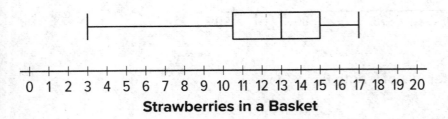

**Strawberries in a Basket**

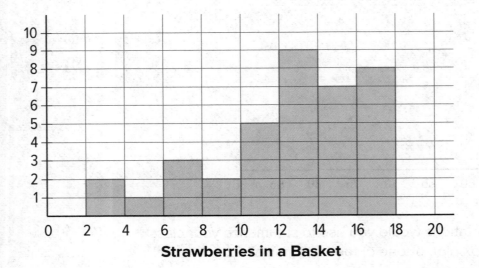

**Strawberries in a Basket**

NAME _____ DATE _____ PERIOD _____

1. Kiran makes these claims. For each claim, decide whether you agree or disagree. Explain your reasoning using at least one of the data displays.

    a. There are 4 baskets that contain 11 strawberries.

    b. The range of the number of strawberries in baskets can be found using any of the three data displays.

    c. The number of baskets of strawberries can only be found using the dot plot.

    d. The interquartile range can be found using the dot plot or box plot, but is easiest with the box plot.

    e. The total number of strawberries in all the baskets can only be determined from the dot plot.

**2.** Complete the table to show the frequency of baskets containing strawberries in each range. Which representation did you use?

| Number of Strawberries | Frequency |
|---|---|
| 0–6 | |
| 6–12 | |
| 12–18 | |

**Lesson 1-4**

# The Shape of Data Distributions

NAME _____ DATE _____ PERIOD _____

**Learning Goal** Let's explore various shapes of data.

## Warm Up
### 4.1 Math Talk: Number Line Distance

Mentally, find the distance between the two values on a number line.

- 70 and 62
- 70 and 70
- 70 and 79
- 70 and 97

## Activity
### 4.2 Suspicious Descriptions

For each picture and description:

- Do you agree or disagree with the description?
- If you agree, explain how you know it is correct.
- If you disagree, explain the error and write the correct description. Explain how you know it is correct.

Bell-shaped since there is a central peak for symmetric data that is less frequent on the ends.

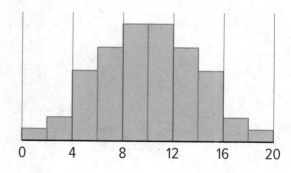

Symmetric because if the distribution was cut in half, both sides would be the same shape.

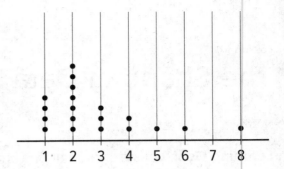

Uniform because there seems to be the same amount of data points across the entire distribution.

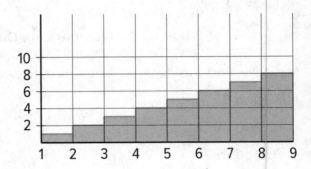

Symmetric because if the distribution was cut in half, both sides would be the same shape.

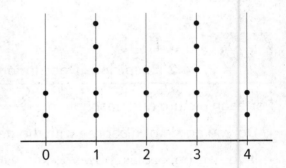

NAME _____ DATE _____ PERIOD _____

Skewed left since most of the data is on the left side of the distribution.

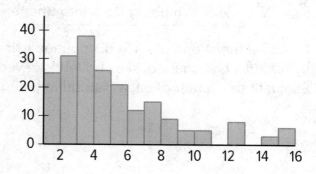

### 4.3 Whipping Data into Shape

Describe the shape of each distribution using the terms approximately, symmetric, bell-shaped, skewed left, skewed right, uniform, or bimodal. Estimate the center of each distribution.

**A**

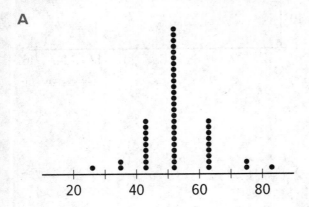

**B**

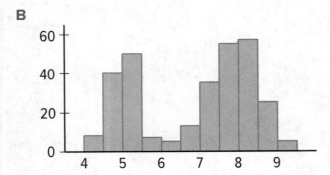

**C**

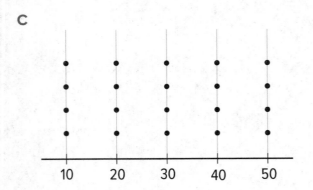

NAME _____ DATE _____ PERIOD _____

**D**

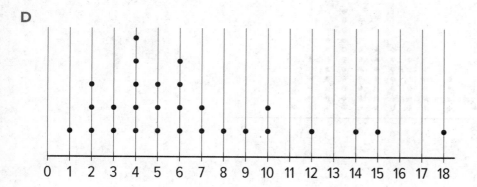

**E**

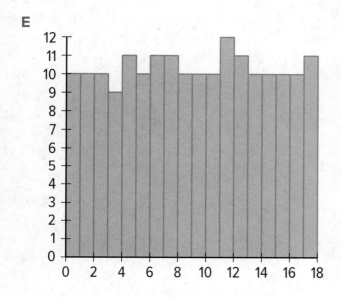

**F**

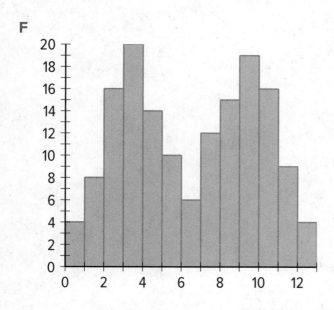

**G**

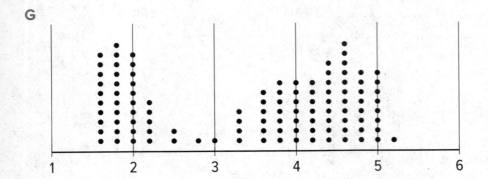

**H**

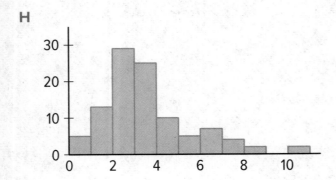

Lesson 1-5

# Watch Your Calculations

NAME _____ DATE _____ PERIOD _____

**Learning Goal** Let's calculate mean, median, mean absolute deviation (MAD), and interquartile range

 ## Warm Up

### 5.1 Math Talk: Find the Mean

Find the mean of each data set.

40, 40, 40, 44

40, 40, 40, 36

40, 40, 40, 100

40, 40, 40, 0

The high school principal wants to know which tenth grade students can be enrolled in an advanced literature course based on their current reading scores. The reading scores are on a scale of 0–800, with a score of 490 considered qualified for the advanced literature course. Here are the students' scores:

| 500 | 525 | 520 | 525 | 525 | 500 | 500 | 520 |
|-----|-----|-----|-----|-----|-----|-----|-----|
| 520 | 500 | 230 | 270 | 200 | 300 | 300 | 300 |
| 315 | 345 | 345 | 400 | 400 | 400 | 450 | 450 |
| 470 | 515 | 550 | 550 | 550 | 600 | 600 | 625 |
| 625 | 700 | 720 | 720 | 800 | 600 | 200 | |

List 1 is a list of measures of center and measures of variability, and List 2 describes the steps you take to calculate the measures.

- Match each measure from List 1 with the way it is computed in List 2.

- Compute each measure for the given list of reading scores.

- Which measures tell you about the center of the data, and which tell you about the variability?

NAME _____ DATE _____ PERIOD _____

**List 1:**

1. Median

2. Interquartile range

3. Mean absolute deviation

4. Mean

**List 2:**

A. Add up all of the values in a data set, then divide by the number of values in the set.

B. The difference between the first and the third quartiles.

C. Find the distance between the mean and each value in the data set. Then, find the mean of those distances.

D. List the values in the data set in order, then find the middle value. If there are two "middle values," find the mean of those two values.

Based on the values, would you say the class is qualified for advanced literature? How does the variability affect your answer?

## Activity

### 5.3 Row Game: Calculations

Work independently on your column. Partner A completes the questions in column A only and partner B completes the questions in column B only. Your answers in each row should match. Work on one row at a time and check if your answer matches your partner's before moving on. If you don't get the same answer, work together to find any mistakes.

| Row | Column A | Column B |
|-----|----------|----------|
| 1 | Calculate the mean<br>1, 1, 1, 2, 100 | Calculate the mean<br>20, 20, 20, 20, 25 |
| 2 | Calculate the mean<br>90, 86, 82, 83.5, 87 | Calculate the mean<br>100, 96, 93.5, 90, 49 |
| 3 | Find the median<br>9, 4, 10, 1, 6 | Find the median<br>6, 11, 12, 2, 4 |
| 4 | Calculate the IQR<br>13, 20, 12, 14, 19, 17, 11, 15, 16 | Calculate the IQR<br>2, 5, 8, 9, 1, 3, 10, 4, 6 |
| 5 | Calculate the IQR<br>55, 50, 52, 49, 34, 36, 40, 46 | Calculate the IQR<br>40, 43, 52, 50, 30, 36, 42, 59 |
| 6 | Calculate the MAD<br>1.75, 2.20, 2.50, 2.55, 2.75, 2.80, 3.00, 4.45 | Calculate the MAD<br>2.75, 3.20, 3.50, 3.55, 3.75, 3.80, 4.00, 5.45 |

Lesson 1-9

# Using Technology for Statistics

NAME _____ DATE _____ PERIOD _____

**Learning Goal** Let's explore technology used to represent data and calculate statistics.

## Warm Up

### 9.1 Estimation: Stack of Books

How tall is the stack of books?

1. Record an estimate that is:

| Too Low | About Right | Too High |
|---------|-------------|----------|
|         |             |          |

2. Explain your reasoning.

## Activity

### 9.2 Spreadsheet Statistics

Here is a list of the number of pages for fiction books on a shelf.

| | | | | | | | |
|-----|-----|-----|-----|-----|-----|-----|-----|
| 15  | 243 | 426 | 175 | 347 | 186 | 236 | 394 |
| 170 | 412 | 242 | 479 | 185 | 254 | 186 | 278 |
| 277 | 278 | 486 | 207 | 378 | 251 | 458 | 360 |
| 440 | 181 | 349 | 482 | 382 |     |     |     |

1. Describe a method to find the mean of these values. (Do not calculate the value yet.)

2. Input the data into a spreadsheet in the same column. In another column, write a spreadsheet formula that will compute the value of the mean for the values in the column. What is the mean number of pages in these books?

3. The first book on the list with 15 pages was a recording error and should have been 158 pages. You need to compute the new mean using 158 instead of 15. Would you rather compute the new mean by hand or use a computer? Explain your reasoning.

4. What is the new mean number of pages in these books with the updated value of 158 pages for the first book in the list?

5. Another book is found and added to the shelf. This new book has 519 pages. How would you update your spreadsheet formula to include this new value?

NAME _____ DATE _____ PERIOD _____

## Activity

### 9.3 Which Data Display?

For each set of data, select the data display that is most informative, then explain your reasoning.

The total area of 50 U.S. states in thousands of square kilometers.

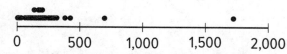

**Area in Thousands of Square Kilometers**

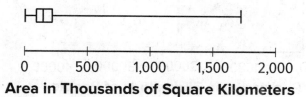

**Area in Thousands of Square Kilometers**

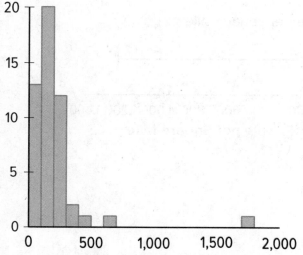

**Area in Thousands of Square Kilometers**

The population of the 50 U.S. states in thousands of people.

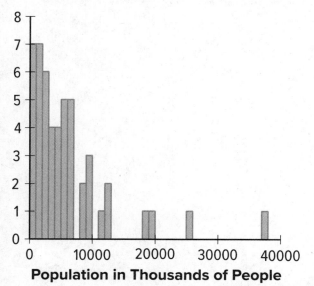

**Population in Thousands of People**

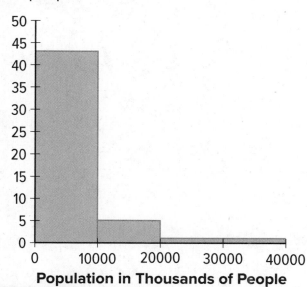

**Population in Thousands of People**

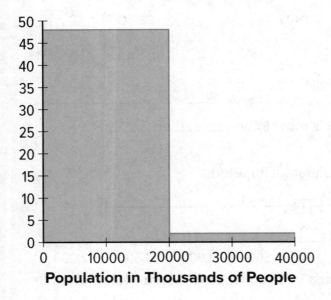

**Population in Thousands of People**

The population density of 50 U.S. states in people per square mile.

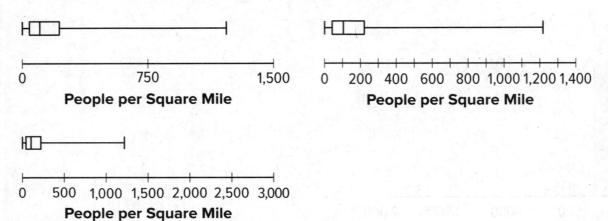

**People per Square Mile**

**People per Square Mile**

**People per Square Mile**

**Lesson 1-10**

# Measures of Center

NAME _____ DATE _____ PERIOD _____

**Learning Goal** Let's explore the relationship between measures of center and the shape of data.

 **Warm Up**

**10.1 Estimation: Lamp Post**

How tall is the lamp post?

1. Record an estimate that is:

| Too Low | About Right | Too High |
|---------|-------------|----------|
|         |             |          |

2. Explain your reasoning.

For each graph, estimate the balance point. The balance point is where you think the number line would balance. Record your balance point on the graph with a △ symbol. Then, calculate the mean and median for each data set.

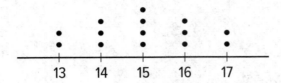

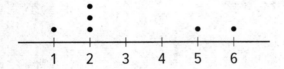

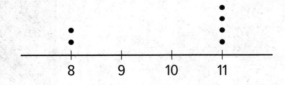

NAME _____ DATE _____ PERIOD _____

# Activity
## 10.3 Mean vs. Median

When people join a gym, they are assessed on their fitness by doing several exercises. The results are given as a score between 1 and 100 with 100 representing peak fitness for the person's age. The gym claims they can improve scores for members after only 2 months.

After 2 months, 11 people are assessed again, and the number of points they improve on in the assessment is shown in the dot plot.

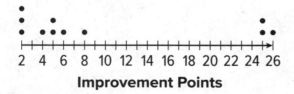

**Improvement Points**

1. What is the mean improvement among these members?

2. What is the median?

3. Which measure of center is a better representation of the members' improvement? Explain your reasoning.

# Decisions, Decisions

NAME _____ DATE _____ PERIOD _____

**Learning Goal** Let's compare data sets

## Warm Up
### 11.1 Estimation: Stack of Books II

How tall is the stack of books?

**1.** Record an estimate that is:

| Too Low | About Right | Too High |
|---------|-------------|----------|
|         |             |          |

**2.** Explain your reasoning.

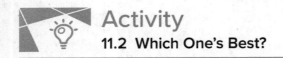

## Activity
### 11.2 Which One's Best?

In order from least to greatest, here are the prices per gallon of gas at two different gas stations over the past 7 days.

Station A:

2.38, 2.68, 2.82, 2.86, 2.99, 3.26, 3.59

Station B:

2.84, 2.85, 2.88, 2.95, 2.98, 3.03, 3.05

Suppose that these gas stations were the closest to your house, but not near each other. Which gas station would you go to for gas? Explain your reasoning.

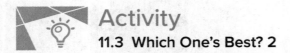

## Activity
### 11.3 Which One's Best? 2

For each pair of data, decide which one you would choose. Use the median and interquartile range to support your choice.

1. A family is trying to decide which restaurant to go to. Here are each restaurant's health inspection ratings over the past year. Based on the restaurants' ratings, which restaurant should the family go to?

   a. Restaurant A: 88, 87, 89, 90, 87, 85, 88, 91, 86, 86, 88, 89

   b. Restaurant B: 90, 65, 89, 50, 94, 93, 95, 95, 75, 70, 88, 89

2. At the end of last year, teachers were rated by their students on a 0–10 scale. Two of the teachers' ratings are given. Whose class would you register for? Explain your reasoning.

   a. Teacher A: 9, 8, 10, 10, 7, 1, 8, 1, 2, 8

   b. Teacher B: 9, 8, 8, 7, 9, 7, 7, 9, 7, 8

Lesson 1-12

# Variability

NAME _____ DATE _____ PERIOD _____

**Learning Goal** Let's understand variability in situations.

 ## Warm Up

### 12.1 Estimation: Average Book Length

The book in the center is 276 pages long. What is the mean number of pages of the 3 books?

1. Record an estimate that is:

| Too Low | About Right | Too High |
|---------|-------------|----------|
|         |             |          |

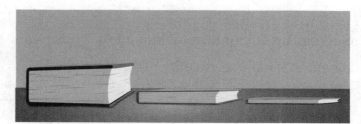

2. Explain your reasoning.

Choose the best option based on the given data. Use a measure of center and measure of variability to justify your answers.

Participants' ages at camp A

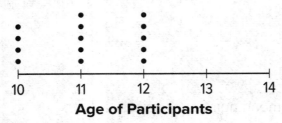

**Age of Participants**

Participants' ages at camp B

mean: 11.93 years, mean absolute deviation: 1.65 years

Servers' tip amounts at restaurant A

mean: $15, mean absolute deviation: $1.5

Servers' tip amounts at restaurant B

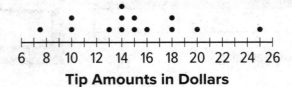

**Tip Amounts in Dollars**

NAME _____ DATE _____ PERIOD _____

1. Which camp would you prefer to work for? Explain your reasoning.

2. Which camp would you prefer to attend? Explain your reasoning.

3. At which restaurant would you want to be a server? Explain your reasoning.

# Activity

## 12.3 Notice and Wonder: Preschool Heights

Mai and Tyler both visit the same preschool classroom and measure the heights of people in the room in inches. The summary of their results are shown in the box plot and statistics.

What do you notice? What do you wonder?

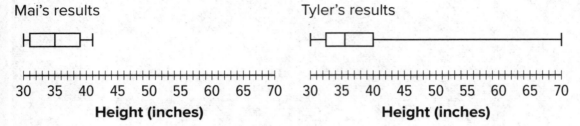

Mai's results

Tyler's results

Height (inches)

Height (inches)

mean: 35.2 inches, mean absolute deviation: 3.24 inches

mean: 39.3 inches, mean absolute deviation: 7.19 inches

Lesson 1-13

# Standard Deviation in Real-World Contexts

NAME _____ DATE _____ PERIOD _____

**Learning Goal** Let's think about standard deviation in the real world.

## Warm Up

### 13.1 Estimation: Marathon Runner

How long will it take the runner to finish the marathon?

1. Record an estimate that is:

| Too Low | About Right | Too High |
|---------|-------------|----------|
|         |             |          |

2. Explain your reasoning.

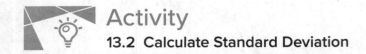

Movie A ratings on a 10 point scale:

9, 8, 10, 10, 7, 1, 8, 1, 2, 8

Restaurant A ratings on a 100 point scale:

88, 87, 89, 90, 87, 85, 88, 91, 86, 86, 88, 89

Movie B ratings on a 10 point scale:

9, 8, 8, 7, 9, 7, 7, 9, 7, 8

Restaurant B ratings on a 100 point scale:

90, 65, 89, 50, 94, 93, 95, 95, 75, 70, 88, 89

1. Calculate the mean and standard deviation for each data set.

NAME _____ DATE _____ PERIOD _____

2. Based on these statistics, which movie and restaurant would you choose? Explain your reasoning.

 Activity

### 13.3 Which Route is the Best Route?

Priya timed the ride from home to school on two different routes.

Here are the times in minutes:

Route A: 21.5, 23, 24, 25, 26.5          Route B: 12, 20, 24, 28, 36

1. Before calculating the standard deviation, predict which route has a greater standard deviation. Explain your reasoning.

**2.** Calculate the standard deviation and use it to decide which route you would recommend for Priya.

**Lesson 1-14**

# Outliers & Means

NAME _____ DATE _____ PERIOD _____

**Learning Goal** Let's explore outliers in data sets.

## Warm Up
### 14.1 Math Talk: Outliers

Solve each expression mentally.

$20 + 1.5(10)$

$20 - 1.5(10)$

$20 + 1.5(14)$

$20 + 1.5(13)$

## Activity

### 14.2 Mountain Hike

Andre records how long it takes him (in minutes) to hike a mountain each day for 6 days.

50     52     58     55     59     50

1. Calculate the mean number of minutes it takes Andre to hike a mountain.

2. Andre plans to hike the same mountain trail one more day. Estimate the time it will take him to complete the trail for the seventh day. Explain your reasoning.

3. What do you think will happen to the mean time for the week if Andre's grandfather comes on the hike with him for the seventh day?

4. Calculate the mean number of minutes using the values when Andre's grandfather comes on the hike for the seventh day: 50, 52, 58, 55, 59, 50, 130.

5. If Andre's grandfather did not come with him on the hike, Andre thinks he could have finished the trail in 60 minutes. Calculate the mean hiking time using Andre's estimate for the seventh day: 50, 52, 58, 55, 59, 50, 60.

NAME _____ DATE _____ PERIOD _____

 ## Activity

### 14.3 The Meaning of an Outlier

For each set of data, answer these questions:

- Use the data to compute the quartiles (Q1 and Q3) and the interquartile range.

- Use the expressions $Q1 - 1.5 \cdot IQR$ and $Q3 + 1.5 \cdot IQR$ to help find where an outlier may be.

- Are any of the values outliers? Explain your reasoning.

1. A group of students recorded the distance, in miles, of the park nearest their home:

   2.3, 4, 1.6, 15, 3.8, 0.75, 1.7

2. Han visits a website to price the next phone he wants to get. He sees the following prices, in dollars:

   200, 485, 492, 512, 453, 503

3. The amount of points Clare scored in her last 8 basketball games are:

   17, 14, 16, 2, 13, 14, 15, 17

4. Kiran's math test scores, as a percentage, were:

   57, 82, 80, 85, 89, 84

5. The height in feet of the roller coasters at the amusement park are:

   415, 456, 423, 442, 30

Lesson 1-15

# Where Are We Eating?

NAME _____ DATE _____ PERIOD _____

**Learning Goal** Let's use graphical representations to compare real-world scenarios.

 ## Warm Up
### 15.1 Estimation: Marathon Runner 2

How long will it take the marathon runner to finish the marathon?

**1.** Record an estimate that is:

| Too Low | About Right | Too Low |
|---------|-------------|---------|
|         |             |         |

**2** Explain your reasoning.

These distributions represent marathon times for different groups.

A

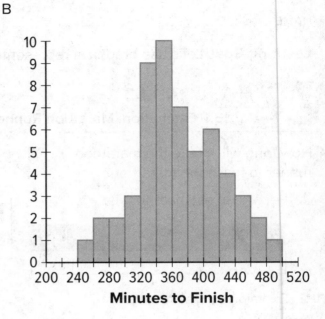

B

C

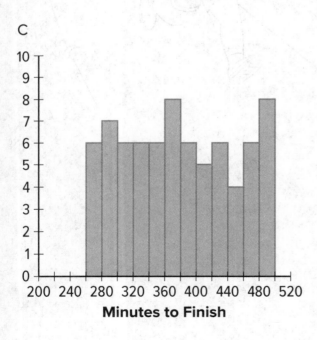

NAME _____  DATE _____  PERIOD _____

1. Which display is most likely to represent the marathon times for people aged 20–30? Explain your reasoning.

2. Which display is most likely to represent the marathon times for every tenth person to cross the finish line? Explain your reasoning.

3. Which display is most likely to represent the marathon times for people aged 40–50? Explain your reasoning.

## Activity

### 15.3  Where Are We Eating?

Restaurant A

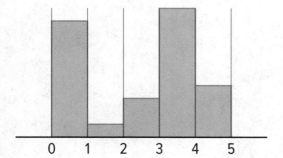

Restaurant B

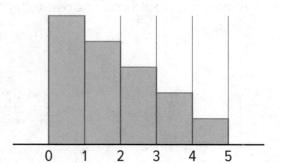

Restaurant C

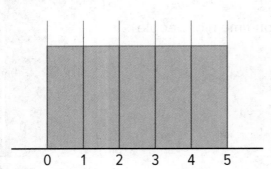

Restaurant D

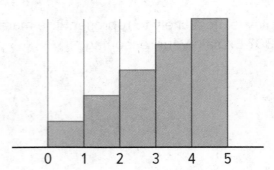

Each histogram represents the number of star ratings for a different restaurant. The ratings range from 0–4, with 0 representing a very poor experience and 4 representing an excellent experience. For each question, explain your reasoning.

1. Which restaurant do reviewers like the most?

2. Which restaurant do reviewers like the least?

3. Which restaurant received mostly mixed reviews?

4. Which restaurant would you choose to try?

Lesson 1-16

# Compare & Contrast

NAME _____ DATE _____ PERIOD _____

**Learning Goal** Let's analyze data.

## Warm Up
### 16.1 Math Talk: Measuring Up

What is the distance between the markings?

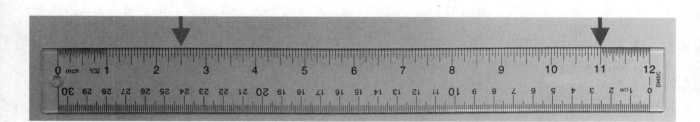

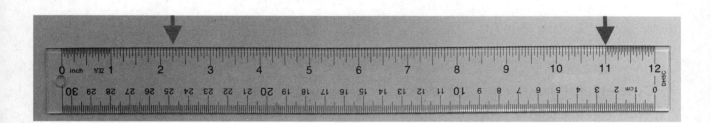

Illustrative Math

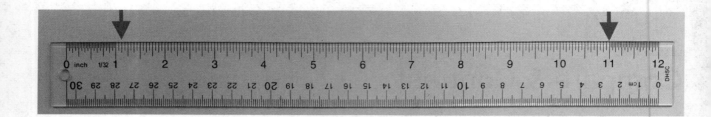

## Activity

### 16.2 Compare & Contrast

Here are the shoe sizes from two cohorts in the military.

Cohort A:

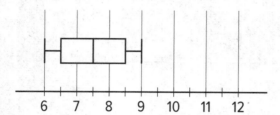

Cohort B:

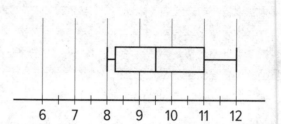

1. Is there any overlap between the two data sets? Explain your reasoning.

2. Which cohort has more variability?

3. Does at least one person from cohort A have a bigger shoe size than someone from cohort B? Explain your reasoning.

4. Compare the measures of center.

NAME _____ DATE _____ PERIOD _____

## Activity
### 16.3 Compare & Contrast Continued

Here are the shoes sizes of some grade 9 and grade 12 students.

Grade 9 shoe sizes:

    6    8   6.5  7.5   7   6.5   9    6
   8.5  7.5   8   10   11    8    9

Grade 12 shoe sizes:

   10    9   10.5  8.5   10    9   9.5   8
    8   11    9   9.5   11   10.5  8.5

1. Create a box plot, dot plot, or histogram to represent both sets of data.

**2.** Describe the distribution shapes.

**3.** Complete the table.

| | Mean | Median | IQR | Standard Deviation |
|---|---|---|---|---|
| Grade 9 Shoe Sizes | | | | |
| Grade 12 Shoe Sizes | | | | |

**4.** Does one grade's shoe sizes have more variation than the other? Explain how you know.

**5.** Compare the measures of center for the two sets of shoe sizes.

**6.** Do the distributions overlap? Use the data display you created to explain how you know.

# Linear Equations, Inequalities, and Systems

The volunteers who clean up trash after a parade could use linear equations to determine the most efficient route for trash collection. You will learn more about linear systems in this unit.

## Topics

- Writing and Modeling with Equations
- Manipulating Equations and Understanding Their Structure
- Systems of Linear Equations in Two Variables
- Linear Inequalities in One Variable
- Linear Inequalities in Two Variables
- Systems of Linear Inequalities in Two Variables

# Linear Equations, Inequalities, and Systems

Lesson 2-1

# Expressing Mathematics

NAME _____ DATE _____ PERIOD _____

**Learning Goal** Let's use operations and variables to describe situations.

 ## Warm Up
### 1.1 Notice and Wonder: Party Planning

What do you notice? What do you wonder?

"Kiran is helping his aunt and uncle plan a cookout. His family has a lot of experience planning parties."

"Kiran's uncle is in charge of the food. He tells Kiran that he plans to use $\frac{1}{4}$ pounds of ground beef per person and 2 ears of corn per person."

"Kiran's aunt is getting plates and paper towels. She plans on buying one plate per person, plus 10 extra plates just in case, and she's going to buy one roll of paper towels for every 10 people."

## Activity

### 1.2 Feeding Operation

A zookeeper is preparing to care for snakes in an exhibit. For each question, write an expression representing the supplies needed.

1. She needs one mouse for each snake, plus two extra mice. How many mice are needed if the number of the snakes is:

   a. 10

   b. 6

   c. $x$

2. She needs 4.5 ounces of crickets for each snake. How many ounces of crickets are needed if the number of snakes is:

   a. 10

   b. 6

   c. $x$

NAME _____ DATE _____ PERIOD _____

**3.** For every 2 snakes, she needs 1 bowl of water. How many bowls of water are needed if the number of snakes is:

   **a.** 10

   **b.** 6

   **c.** $x$

**4.** There is one male snake, and the rest are female. She needs one vitamin pill for every female snake. How many vitamin pills does she need if the number of snakes is:

   **a.** 10

   **b.** 6

   **c.** $x$

## Activity
### 1.3 Important Quantities

To understand the situation, what is some information you would like to know? What information is already given?

1. Clare is in charge of getting snacks for a road trip with her two friends and her dog. She has $35 to go to the store to get some supplies.

2. Andre wants to surprise his neighbor with a picnic basket of fruit and vegetables from his garden. The basket can hold up to 12 food items.

3. Tyler is packing his bags for vacation.

4. Mai's teacher orders tickets to the local carnival for herself, the entire class, and 2 more chaperones.

5. Jada wants to prepare the fabric for the bridesmaids dresses she is creating for a wedding party.

Lesson 2-2

# Words and Symbols

NAME _____ DATE _____ PERIOD _____

**Learning Goal** Let's use symbols to represent words in a math problem

## Warm Up
### 2.1 Math Talk: Perceiving Percents

Evaluate mentally:

50% of 80                                     1% of 80

75% of 80                                     7.5% of 80

## Activity
### 2.2 Identify and Represent

For each problem, identify any important quantities. If it's a known quantity, write the number and a short description of what it represents. If it's an unknown quantity, assign a variable to represent it and write a short description of what that variable represents.

1. Clare is in charge of getting snacks for a road trip with her friends and her dog. She has $35 to go to the store to get some supplies. The snacks for herself and her friends cost $3.25 each, and her dog's snacks costs $9 each.

**2.** Tyler is packing his bags for vacation. He plans to pack two outfits for each day of vacation.

**3.** Mai's teacher orders tickets to the local carnival for herself, the entire class, and 3 more chaperones. Student tickets are $4.50.

**4.** Jada wants to prepare the fabric for the bridesmaids' dresses she is creating for a wedding party. She plans to use about 16 square feet of fabric on each dress.

**5.** Elena is going to mow lawns for the summer to make some extra money. She will charge $20 for every lawn she mows and plans on mowing several lawns each week.

## Activity
### 2.3 Matching Expressions

Your teacher will give you a set of cards. Group them into pairs by matching each situation with an algebraic expression. Some cards do not have matches. Use the blank cards to create your own expression or situation so that all cards have an accurate match.

**Lesson 2-3**

# Setting the Table

NAME _____ DATE _____ PERIOD _____

**Learning Goal** Let's look at different ways to represent the same relationships.
Let's look closely at tables.

 ## Warm Up
**3.1 Notice and Wonder: A Table**

What do you notice? What do you wonder?

| x | y |
|---|---|
| 0 | 6 |
| 1 | 9 |
| 2 | 12 |
| 4 | 18 |
| 10 | 36 |
| 100 | |

## Activity
### 3.2 Complete the Table

Complete the table so that each pair of numbers makes the equation true.

1. $y = 3x$

| x | y |
|---|---|
| 5 | |
| | 96 |
| $\frac{2}{3}$ | |

2. $m = 2n + 1$

| n | m |
|---|---|
| 3 | |
| | 5 |
| | 12 |

NAME _____ DATE _____ PERIOD _____

**3.** $s = \dfrac{t-1}{3}$

| t | s |
|---|---|
| 0 | |
| | 4 |
| | 52 |

**4.** $d = \dfrac{16}{e}$

| e | d |
|---|---|
| 4 | |
| -3 | |
| | 2 |

## Activity

### 3.3 Card Sort: Tables, Equations, and Situations

1. Take turns with your partner to match a table, a situation, and an equation. On your turn, you only need to talk about two cards, but eventually all the cards will be sorted into groups of 3 cards.

2. For each match that you find, explain to your partner how you know it's a match. Ask your partner if they agree with your thinking.

3. For each match that your partner finds, listen carefully to their explanation. If you disagree, discuss your thinking and work to reach an agreement.

Lesson 2-4

# Solutions in Context

NAME _____ DATE _____ PERIOD _____

**Learning Goal** Let's use equations to describe situations.

## Warm Up
### 4.1 Notice and Wonder: Equations

What do you notice? What do you wonder?

- $2x + 3y = 12$

- $(0, 4)$ and $(6, 0)$

## Activity
### 4.2 Raffles and Snacks

1. For a fundraiser, a school club is selling raffle tickets for $2 each and healthy snacks for $1.50 each. What is the cost of:

   a. 3 tickets?

   b. 5 tickets?

   c. $x$ tickets?

   d. 2 snacks?

   e. 6 snacks?

   f. $y$ snacks?

   g. 10 tickets and 8 snacks?

   h. 7 tickets and 5 snacks?

   i. $x$ tickets and $y$ snacks?

NAME _____ DATE _____ PERIOD _____

**2.** Lin bought some tickets and some snacks, and paid $22.

　　**a.** Write an equation representing this situation.

　　**b.** What are some combinations of tickets and snacks that Lin might have purchased?

## Activity

### 4.3 Row Game: Solving Equations

Partner A completes only column A, and partner B completes only column B. Your answers for each problem should match. Work on one problem at a time, and check whether your answer matches your partner's before moving on. If you don't get the same answer, work together to find your mistake.

| Column A | Column B |
|---|---|
| 1. Lin's teacher has a daughter that is $\frac{1}{3}$ of his age. Write an expression to represent the daughter's age. Let $z$ represent the teacher's age, in years. | 1. Jada leaves the beach with some seashells. One out of every three of the shells turns out to contain a hermit crab. Write an equation to represent the number of hermit crabs Jada found. Let $z$ represent the total number of seashells she collected. |
| 2. Han wants to save $40. He hasn't met his goal yet. Write an expression to represent how far Han is from his goal, in dollars. Let $q$ represent the amount of money, in dollars, he's saved so far. | 2. Tyler started the school year with 40 pencils, but he's lost some. Write an equation to represent how many pencils Tyler has left. Let $q$ represent the number of pencils he's lost so far. |
| 3. Priya has some money to spend at a fair. It costs $6 to get in and $.50 per ride ticket. Write an expression to represent how much Priya spends at the fair, in dollars. Let $x$ represent the number of ride tickets Priya buys. | 3. When Clare bought her plant, it was 6 inches tall. Each week, it's been growing $\frac{1}{2}$ of an inch. Write an expression to represent how tall Clare's plant is, in inches. Let $x$ represent the number of weeks since Clare bought her plant. |
| 4. Diego is inviting some friends over to watch movies. He is buying popcorn and peanuts. Popcorn costs 6 cents per ounce and peanuts cost 17 cents per ounce. Write an expression to represent the total cost of peanuts and popcorn, in cents. Let $j$ represent how many ounces of popcorn Diego buys and $k$ represent how many ounces of peanuts he buys. | 4. Mai is packing care packages. She is putting in boxes of granola bars that weigh 6 ounces each and paperback books that weigh 17 ounces each. Write an expression to represent the total weight of a care package, in ounces. Let $j$ represent the number of boxes of granola bars and $k$ represent the number of books. |

**Lesson 2-5**

# Graphs, Tables, and Equations

NAME _____ DATE _____ PERIOD _____

**Learning Goal** Let's connect different representations.

 Warm Up

**5.1 Math Talk: Solving Equations**

Solve each equation mentally.

$100 = 10(x - 5)$

$300 = 30(x - 5)$

$15 - 971 = x - 4 - 971$

$\frac{10}{7} = \frac{1}{7}(x - 19)$

## Activity

### 5.2 On the Line

1. Sketch a graph representing each of these equations.

   a. $y = 2x$

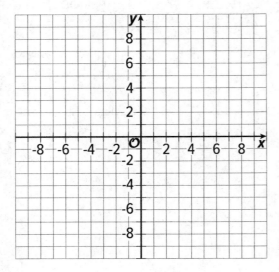

   b. $y = \frac{1}{2}x$

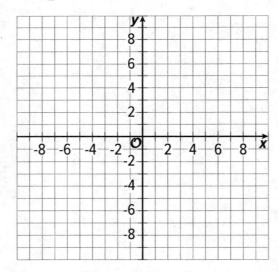

NAME _____ DATE _____ PERIOD _____

c. $y = x + 2$

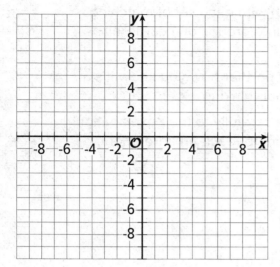

2. For each point, which graph or graphs is it on? How can you tell by using the equation?

   a. (1, 3)

   b. (0, 0)

   c. (3, 6)

   d. (3, 1.5)

## Activity

### 5.3 Take Turns: Graphs, Tables, Equations, and Situations

1. Take turns with your partner to match a graph with each set of matching cards. Eventually all the cards will be sorted into groups of 4 cards (an equation, situation, table, and graph).

2. For each match that you find, explain to your partner how you know it's a match. Ask your partner if they agree with your thinking.

3. For each match that your partner finds, listen carefully to their explanation. If you disagree, discuss your thinking and work to reach an agreement.

Lesson 2-6

# Equality Diagrams

NAME _____ DATE _____ PERIOD _____

**Learning Goal** Let's use hanger diagrams to understand equivalent equations.

## Warm Up
### 6.1 Notice and Wonder: Solving Equations

What do you notice? What do you wonder?

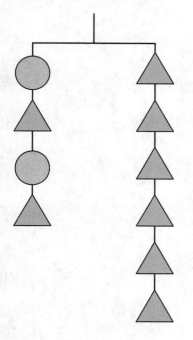

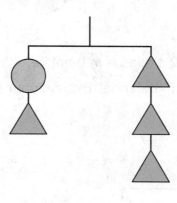

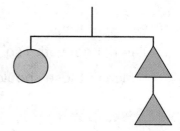

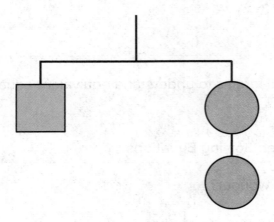

1. The hanger with 1 square and 2 circles is in balance.

   Which of these should also be in balance? Explain your reasoning.

(A.)

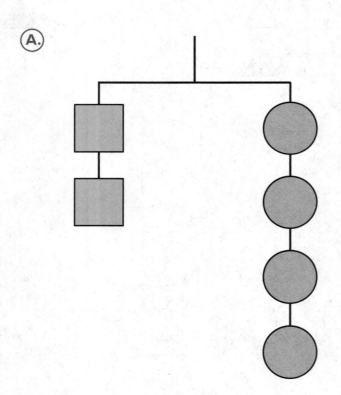

NAME _____ DATE _____ PERIOD _____

**B.**

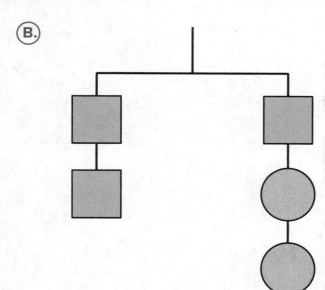

**C.**

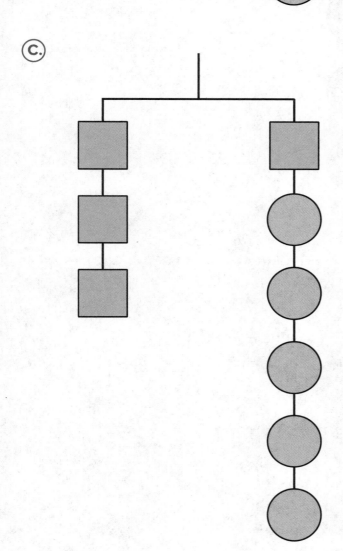

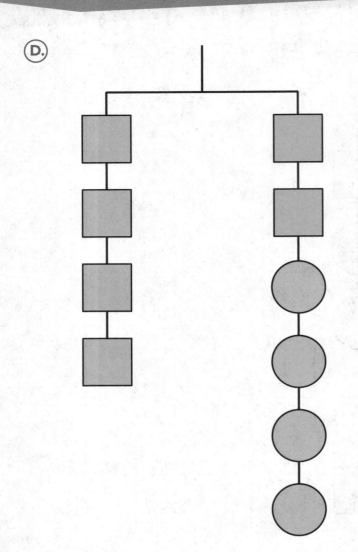

NAME _____ DATE _____ PERIOD _____

**E.**

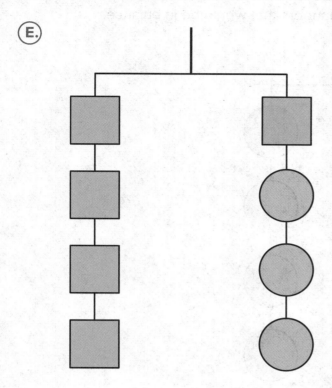

**F.**

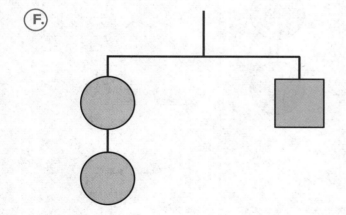

**2.** This hanger containing 2 pentagons and 6 circles is in balance. Use the hanger diagram to create two additional hangers that would be in balance.

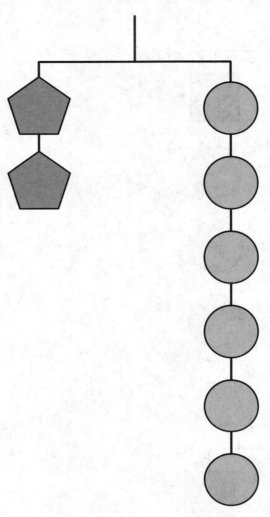

NAME _____ DATE _____ PERIOD _____

## Activity
### 6.3 Diagrams and Equations

In the previous activity, each square weighs 10 pounds and each circle weighs *x* pounds.

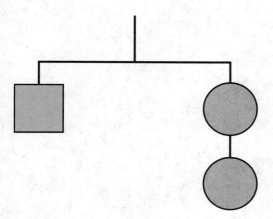

So, this diagram could be represented by the equation. $10 = 2x$

1. Use each of the 6 hanger diagrams containing squares and circles from the previous activity to write an equation that represents the weights on the hanger.

    a.

    b.

    c.

    d.

    e.

    f.

2. Solve each equation.

    a.

    b.

    c.

    d.

    e.

    f.

3. Compare the solutions to the equations with the answers from the previous activity. What do you notice?

Lesson 2-7

# Why Is That Okay?

NAME _____ DATE _____ PERIOD _____

**Learning Goal** Let's rewrite equations without changing their value.

## Warm Up
### 7.1 Estimation: Equal Weights

How many pencils are the same weight as a standard stapler?

**1.** Record an estimate that is:

| Too Low | About Right | Too High |
|---------|-------------|----------|
|         |             |          |

**2.** Explain your reasoning

## Activity

### 7.2  What's the Same? What's Different?

For each pair of equations, decide whether the given value of $x$ is a solution to one or both equations:

1.  Is $x = 2$ a solution to:

    a.  $x(2 + 3) = 10$

    b.  $2x + 3x = 10$

2.  Is $x = 3$ a solution to:

    a.  $x - 4 = 1$

    b.  $4 - x = 1$

3.  Is $x = -2$ a solution to:

    a.  $7x = -14$

    b.  $x \cdot 14 = -28$

4.  Is $x = -1$ a solution to:

    a.  $x + 3 = 2$

    b.  $3 + x = 2$

NAME _____ DATE _____ PERIOD _____

**5.** Is $x = -5$ a solution to:

   **a.** $3 - x = 8$

   **b.** $5 - x = 10$

**6.** Is $x = (8 + 1) + 3$ a solution to:

   **a.** $\dfrac{12}{2} = \dfrac{1}{2}(x)$

   **b.** $18 = 2x$

**7.** Is $x = 2$ a solution to:

   **a.** $\dfrac{12}{x} = 6$

   **b.** $6x = 12$

**8.** Is $x = \dfrac{10}{3}$ a solution to:

   **a.** $-1 + 3x = 9$

   **b.** $9 = 3x - 1$

**9.** Is $x = \dfrac{1}{2}$ a solution to:

   **a.** $5(x + 1) = \dfrac{15}{2}$

   **b.** $5x + 1 = \dfrac{15}{2}$

# Activity

### 7.3 Generating Equivalent Equations

1. Your teacher will display an equation. Take turns with your partner to generate an equivalent equation—an equation with the same solution. Generate as many different equations with the same solution as you can. Keep track of each one you find.

2. For each change that you make, explain to your partner how you know your new equation is equivalent. Ask your partner if they agree with your thinking.

3. For each change that your partner makes, listen carefully to their explanation about why their new equation is equivalent. If you disagree, discuss your thinking and work to reach an agreement.

**Lesson 2-8**

# Reasoning About Equations

NAME _____ DATE _____ PERIOD _____

**Learning Goal** Let's justify the steps of solving an equation.

 ## Warm Up
### 8.1 Math Talk: Multiplying

Evaluate mentally.

$3 \cdot 10$

$3 \cdot 13$

Apply the distributive property.

$3(13 + x)$

$3x(8 - y)$

1. Here is Diego's work.

$$\frac{(4x + 1)}{5} = 2x$$

$$5 \cdot \frac{(4x + 1)}{5} = 5 \cdot 2x$$

$$4x + 1 = 10x$$

$$4x + 1 - 4x = 10x - 4x$$

$$1 = 6x$$

$$\frac{1}{6} \cdot 1 = \frac{1}{6} \cdot 6x$$

$$\frac{1}{6} = x$$

For each step, explain:

a. What did Diego do?

b. Why did Diego do that? How did it help him find the value of $x$ that made the equation true?

c. How could Diego *justify* each move?

NAME _____ DATE _____ PERIOD _____

**2.** Here is an equation and the solution. What moves could you make to get from the equation to the solution? Justify each move you make:

$$12(x - 4) = 2$$
$$x = 4\frac{1}{6}$$

# Activity

## 8.3 Turn Up the Volume

Here are some geometric formulas. In the given problems, you will get some information and be asked to figure out one of the measurements.

As you work, look for patterns or a set of steps that you could use to quickly figure out one measurement, given the others.

Perimeter of a Rectangle: $P = 2l + 2w$     Area of a Rectangle: $A = lw$

Area of a Triangle: $A = \frac{1}{2}bh$        Volume of a Cylinder: $V = \pi r^2 h$

1. Find the missing measurement of the rectangle.

   a. A rectangle has a length of 3.5 units and a width of 9 units. Find its perimeter.

   b. A rectangle has a perimeter of 25 units and a width of 9 units. Find its length.

   c. A rectangle has a perimeter 18 units and a width of 4 units. Find its length.

   d. Look at your steps and answers so far. Are there any patterns you could use to help you solve the next two problems easily?

   e. A rectangle has a perimeter of 24 units and a width of 11 units. Find its length.

   f. A rectangle has a perimeter of 15 units and a width of 3 units. Find its length.

   g. How would you teach someone else to find the length of a rectangle using the patterns you noticed?

NAME _____ DATE _____ PERIOD _____

**2.** Find the missing measurement of the rectangle.

**a.** A rectangle has a length of 4 units and a width of 9 units. Find its area.

**b.** A rectangle has an area of 36 square units and a width of 9 units. Find its length.

**c.** A rectangle has an area 50 square units and a width of 10 units. Find its length.

**d.** Look at your steps and answers so far. Are there any patterns you could use to help you solve the next two problems easily?

**e.** A rectangle has an area of 25 square units and a width of 5 units. Find its length.

**f.** A rectangle has an area of 39 square units and a width of 6 units. Find its length.

**g.** How would you teach someone else to find the length of a rectangle using the patterns you noticed?

**3.** Find the missing measurement of the triangle.

**a.** A triangle has a base of 5 units and a height of 4 units. Find its area.

**b.** A triangle has an area of 10 square units and a height of 4 units. Find its base.

**c.** A triangle has an area of 12 square units and a height of 8 units. Find its base.

**d.** Look at your steps and answers so far. Are there any patterns you could use to help you solve the next two problems easily?

**e.** A triangle has an area of 6 square units and a height of 3 units. Find its base.

**f.** A triangle has an area of 13 square units and a height of 5 units. Find its base.

**g.** How would you teach someone else to find the base of a triangle using the patterns you noticed?

**4.** Find the missing measurement of the cylinder.

   **a.** A cylinder has a height of 3 units and a radius of 5 units. Find its volume.

   **b.** A cylinder has a volume of $75\pi$ cubic units and a radius of 5 units. Find its height.

   **c.** A cylinder has a volume of $90\pi$ cubic units and a radius of 3 units. Find its height.

   **d.** Look at your steps and answers so far. Are there any patterns you could use to help you solve the next two problems easily?

   **e.** A cylinder has a volume of $20\pi$ cubic units and a radius of 2 units. Find its height.

   **f.** A cylinder has a volume of 100 cubic units and a radius of 5 units. Find its height.

   **g.** How would you teach someone else to find the height of a cylinder using the patterns you noticed?

**Lesson 2-9**

# Same Situation, Different Symbols

NAME _____ DATE _____ PERIOD _____

**Learning Goal** Let's think about the how and why of solving equations, and use those ideas to make problems easier.

 ## Warm Up
### 9.1 Math Talk: True Values

For each equation, mentally find the value that makes it true.

- $25 + 3 = 21 + x$
- $5 \cdot 3 + 15 = x \cdot 5 + 10$
- $2 - x + 8 = 2 - 7 + 10$
- $2 \cdot 12 - 50 = 3 \cdot 12 - x$

## Activity

### 9.2 Fizzy Drinks and Fast Driving

1. Sparkling water and grape juice are mixed together to make 36 ounces of fizzy juice.

   a. How much sparkling water was used if the mixture contains 19 ounces of grape juice?

   b. How much grape juice was used if the mixture contains 15 ounces of sparkling water?

   c. Han wrote the equation, $x + y = 36$, with $x$ representing the amount of grape juice used, in ounces, and $y$ representing the amount of sparkling water used, in ounces. Explain why Han's equation matches the story.

   d. Clare wrote the equation $y = 36 - x$, with $x$ representing the amount of grape juice used, in ounces, and $y$ representing the amount of sparkling water used, in ounces. Explain why Clare's equation matches the story.

   e. Kiran wrote the equation $x = y + 36$, with $x$ representing the amount of grape juice used, in ounces, and $y$ representing the amount of sparkling water used, in ounces. Explain why Kiran's equation does *not* match the story.

NAME _____ DATE _____ PERIOD _____

2. A car is going 65 miles per hour down the highway.

   a. How far does it travel in 1.5 hours?

   b. How long does it take the car to travel 130 miles?

   c. Mai wrote the equation $y = 65x$, with $x$ representing the time traveled, in hours, and $y$ representing the distance traveled, in miles. Explain why Mai's equation matches the story.

   d. Tyler wrote the equation $x = \dfrac{y}{65}$, with $x$ representing the time traveled, in hours, and $y$ representing the distance traveled, in miles. Explain why Tyler's equation matches the story.

   e. Lin wrote the equation $y = \dfrac{x}{65}$, with $x$ representing the time traveled, in hours, and $y$ representing the distance traveled, in miles. Explain why Lin's equation does *not* match the story.

**9.3 Finding an Error**

Tyler is practicing finding different equivalent equations that match the story. For each of the problems below, he gets one equation right but the other equation wrong. For each one, explain the error, give the correct equivalent equation, and explain your reasoning.

1. Situation: The yogurt at Sweet Delights costs $0.65 per pound and $0.10 per topping. The total cost of a purchase was $1.70. Let $p$ be the weight of the yogurt in pounds and $t$ be the number of toppings bought.

   Tyler's first and correct equation: $0.65p + 0.10t = 1.70$

   Tyler's second and *incorrect* equation: $t = (1.70 - .65p) \cdot 0.10$

   a. What is the error?

   b. What is a correct second equation Tyler could have written?

   c. What might Tyler have been thinking that led to his mistake?

NAME _____ DATE _____ PERIOD _____

2. Situation: The perimeter of a rectangle (twice the sum of the length and width) is 13.5 inches. Let $\ell$ be the length of the rectangle and $w$ be the width of the rectangle.

Tyler's first and correct equation: $2(\ell + w) = 13.5$

Tyler's second and *incorrect* equation: $w = 13.5 - 2\ell$

a. What is the error?

b. What is a correct second equation Tyler could have written?

c. What might Tyler have been thinking that led to his mistake?

3. Situation: For a fundraiser, a school is selling flavored waters for $2.00 each and pretzels for $1.50 each. The school has a fundraising goal of $200. Let $w$ be the number of waters sold and $p$ be the number of pretzels sold.

   Tyler's first and correct equation: $2w + 1.5p = 200$

   Tyler's second and *incorrect* equation: $1.5p = 198w$

   a. What is the error?

   b. What is a correct second equation Tyler could have written?

   c. What might Tyler have been thinking that led to his mistake?

**Lesson 2-10**

# Equations and Relationships

NAME _____ DATE _____ PERIOD _____

**Learning Goal** Let's match graphs and equations.

 ## Warm Up
### 10.1 Which One Doesn't Belong: Slopes and Intercepts

**A.**

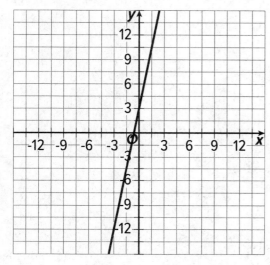

**B.**

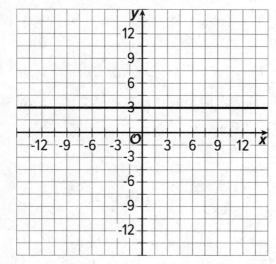

**C.** $y = -2.5x - 7.5$

**D.**

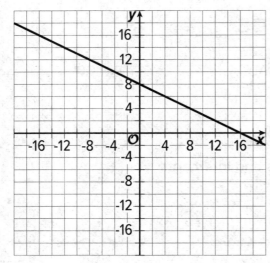

NAME _____ DATE _____ PERIOD _____

## Activity

### 10.2 What's the Same? What's Different?

Here are the graphs of four linear equations.

**Graph A**

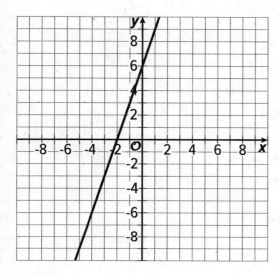

**Graph B**

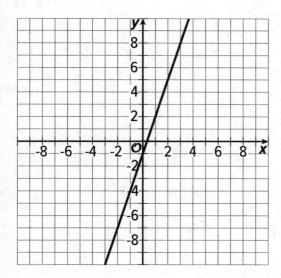

**Graph C**

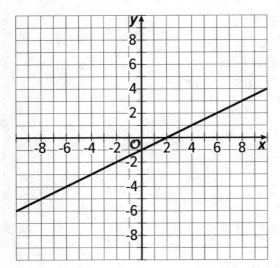

**Graph D**

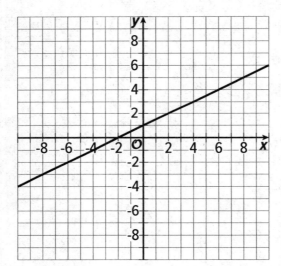

1. Which graphs have a slope of 3?

2. Which graphs have a slope of $\frac{1}{2}$?

3. Which graphs have a $y$-intercept of -1?

4. Which graphs have an $x$-intercept of -2?

NAME _____  DATE _____  PERIOD _____

**5.** Graph A represents the equation $2y - 6x = 12$. Which other equations could graph A represent?

(A.) $y - 3x = 6$

(B.) $y = 3x + 6$

(C.) $y = -3x + 6$

(D.) $2y = -6x + 12$

(E.) $4y - 12x = 12$

(F.) $4y - 12x = 24$

**6.** Write three equations that graph B could represent.

 ## Activity

### 10.3  Situations and Graphs

For each situation, find the slope and intercepts of the graph. Then, describe the meaning of the slope and intercepts. Determine if the values you come up with are reasonable answers for the situation.

**1.** The printing company keeps an inventory of the number of cases of paper it has in stock.

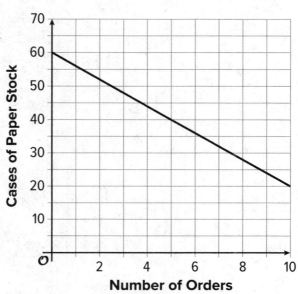

**2.** The market value of a house is determined by the size of the house.

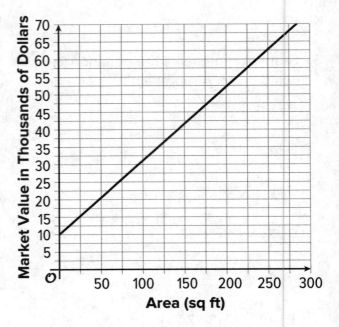

**3.** Tyler teaches paint classes in which the amount of money he makes depends on the number of participants he has.

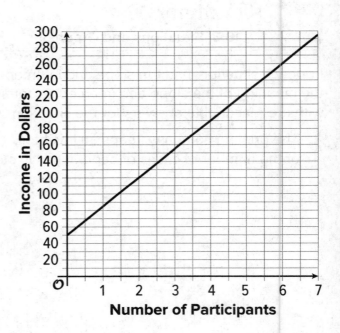

NAME _____ DATE _____ PERIOD _____

4. Mai tracks the amount of money in her no-interest savings account.

5. Priya earns coins for each new level she reaches on her game.

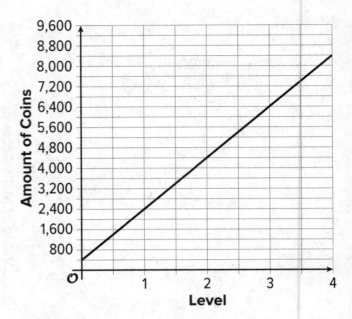

Lesson 2-11

# Slopes and Intercepts

NAME _____ DATE _____ PERIOD _____

**Learning Goal** Let's analyze parts of a graph.

## Warm Up
### 11.1 Notice and Wonder: Sugar and Flour

What do you notice? What do you wonder?

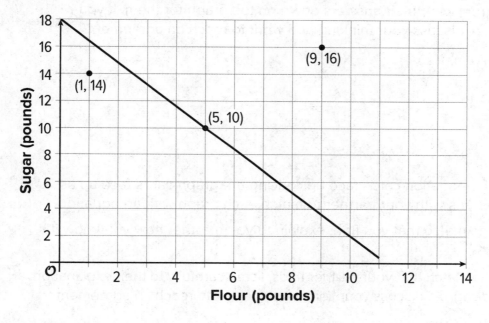

## Activity

### 11.2 Matching Matching

1. Deal out the graph cards, face up. Take turns with your partner to name the three things in each graph:

   - slope
   - *x*-intercept
   - *y*-intercept

   a. For each graph you find, tell your partner the slope, *x*-intercept, and *y*-intercept of each graph.

   b. For each graph your partner finds, see if you agree with their thinking. If you agree, write their answers down on the graph for them. If you disagree, discuss your thinking and work to reach an agreement.

2. Deal out the equation cards face up, leaving the graph cards face up as well. Take turns with your partner to match each graph with an equation.

   a. For each match that you find, explain to your partner how you know it's a match.

   b. For each match that your partner finds, listen carefully to their explanation. If you disagree, discuss your thinking and work to reach an agreement.

## Activity

### 11.3 Part of Linear Equations

For each equation, identify the slope and *y*-intercept of its graph.

1. $y = 3x - 8$

2. $y = 10 - 2x$

3. $y = \frac{x}{2} + 1$

4. $y + 1 = 9x$

5. $y = \frac{1}{3}(9x + 12)$

Lesson 2-12

# Connecting Situations and Graphs

NAME _____ DATE _____ PERIOD _____

**Learning Goal** Let's examine graphs of lines representing situations.

 ## Warm Up
### 12.1 Notice and Wonder: Snacks for Sale

What do you notice? What do you wonder?

A club is selling snacks at a track meet. Oranges cost $1 each and protein bars cost $4 each. They sell a total of 100 items, and collect $304.

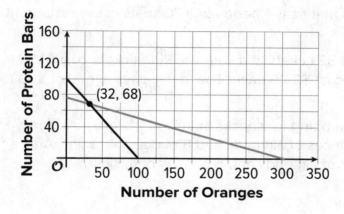

# Activity

## 12.2 Matching Graphs to Situations

1. Match each pair of graphs to a situation.

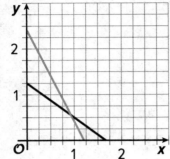

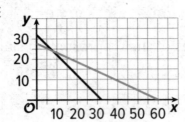

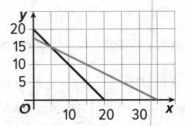

a. A restaurant has a total of 20 tables—round tables that seat 2 people and rectangular tables that seat 4 people. All 70 seats in the restaurant are occupied.

b. A family buys a total of 32 tickets at a carnival. Ride tickets cost $1.50 each and food tickets cost $3.25 each. The family pays a total of $90 for the tickets.

c. Tyler and Andre are shopping for snacks in bulk at the grocery store. Tyler pays $10 for 6 ounces of almonds and 8 ounces of raisins. Andre pays $12 for 10 ounces of almonds and 5 ounces of raisins.

NAME _____ DATE _____ PERIOD _____

**2.** Answer these questions about each situation:

    **a.** What do $x$ and $y$ represent in the situation?

    **b.** At what point do the graphs intersect? What do the coordinates mean in this situation?

## Activity

### 12.3 Ride Sharing Among Friends

A ride sharing company offers two options: riding in small cars that can carry up to 3 passengers each, or riding in large vans that can carry up to 6 passengers each. A group of 27 people is going to use the ride sharing service to take a trip. The trip in a small car costs $10 and the trip in a large van costs $15. The group ends up spending $80 total.

**1.** An equation that represents one of the constraints is $3x + 6y = a$.

    **a.** What is the value of $a$?

    **b.** What do $x$ and $y$ represent?

**2.** An equation that represents the other constraint is $cx + 15y = 80$. What is the value of $c$?

**3.** Here is a graph that represents one of the constraints. Which one? Explain how you know.

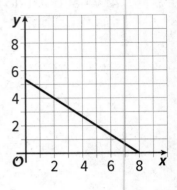

**4.** Sketch another line on the graph that represents the other constraint.

**5.** For each coordinate pair, describe its meaning in the situation and decide whether it satisfies the constraint on total number of people, the constraint on cost, or neither.

  **a.** (2, 4)

  **b.** (1, 4)

  **c.** (3, 2)

**6.** At what point do the two lines intersect? What does this point mean in this situation?

Lesson 2-13

# Making New, True Equations

NAME _____ DATE _____ PERIOD _____

**Learning Goal** Let's practice solving equations.

## Warm Up

### 13.1 Math Talk: Evaluating Expressions

Find the value of $y$ when $x = 5$.

$y = 3x - 4$

$y = \frac{2}{5}x + 4$

$y = 2x + 3 + (3x - 1)$

$y = 4x - (x + 1)$

## Activity

### 13.2 Solving for a Variable

Solve for the indicated variable.

**1.** Solve for $k$. $2t + k = 6$

**2.** Solve for $n$. $10n = 2p$

**3.** Solve for $c$. $12 - 6d = 3c$

**4.** Solve for $g$. $h = 8g + 4$

**5.** Solve for $x$. $4x + 3y = 12$

**6.** Solve for $y$. $4x + 3y = 12$

## Activity

### 13.3 Solving Some Equations

Solve each equation.

| Row | Column A | Column B |
|---|---|---|
| 1 | $4(2x + 8) - 10 = 14$ | $4 + 2(-3x + 5) = 20$ |
| 2 | $3(x - 4) + 6 = 60$ | $3(\frac{1}{2}x + 9) - 5 = 55$ |
| 3 | $4\left(\dfrac{x + 3}{2}\right) - 5 = 10$ | $7 - 2(6x + 1) = -49$ |
| 4 | $2x + (5 - 3x) = 14$ | $1 = 5x + 10 - 4x$ |
| 5 | $4x + 2(3 - x) = 16$ | $x + 2(x - 4) + 5 = 12$ |
| 6 | $2x - 2(3x - 1) = 8$ | $-6x + 2(4x + 5) = 7$ |

Lesson 2-14

# Making More New, True Equations

NAME _____ DATE _____ PERIOD _____

**Learning Goal** Let's practice combining like terms and working with horizontal and vertical lines.

## Warm Up
### 14.1 Criss Cross'll Make You Jump

Match each equation with its graph.

$x = 7$     $y = 7$     $x + y = 7$

A.

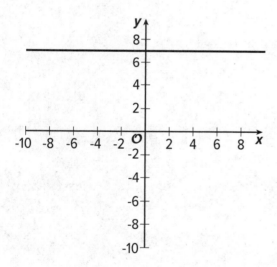

B.

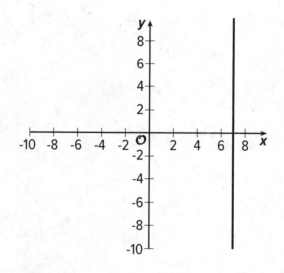

C.

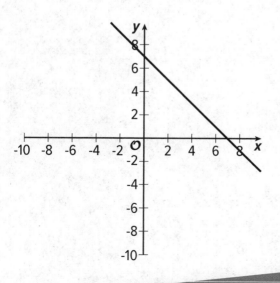

D.

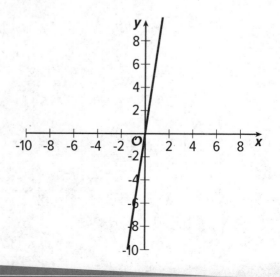

# Activity

## 14.2 They're Like Terms, Man

Rewrite each expression by combining like terms.

1. $11s - 2s$

2. $5t + 3z - 2t$

3. $23s - (13t + 7t)$

4. $7t + 18r + (2r - 5t)$

5. $-4x + 6r - (7x + 2r)$

6. $3(c - 5) + 2c$

7. $8x - 3y + (3y - 5x)$

8. $5x + 4y - (5x + 7y)$

9. $9x - 2y - 3(3x + y)$

10. $6x + 12y + 2(3x - 6y)$

NAME _____ DATE _____ PERIOD _____

## Activity
### 14.3  Finding More Lines

For each system of equations:

- Solve the system of equations by graphing. Write the solution as an ordered pair.

- Write an equation that would be represented by a vertical or horizontal line that also passes through the solution of the system of equations.

- Graph your new equation along with the system.

1. $\begin{cases} y = 3x + 5 \\ y = -x + 1 \end{cases}$

The line representing $y = 3x + 5$ is shown

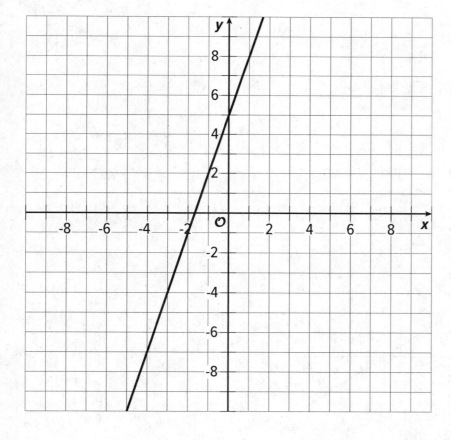

**2.** $\begin{cases} y = \frac{1}{3}x - 2 \\ y = x - 6 \end{cases}$

The line representing $y = \frac{1}{3}x - 2$ is shown

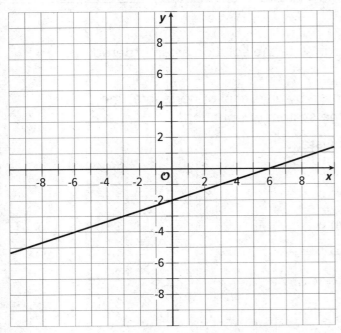

**3.** $\begin{cases} 2x + 3y = 10 \\ x + y = 3 \end{cases}$

The line representing $2x + 3y = 10$ is shown

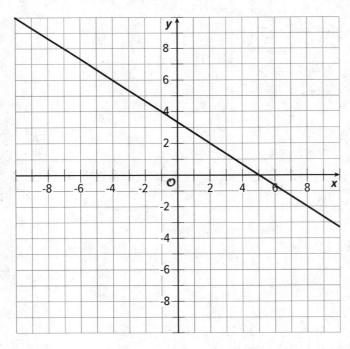

Lesson 2-15

# Off the Line

NAME _____ DATE _____ PERIOD _____

**Learning Goal** Let's study solutions and non-solutions.

 ## Warm Up
### 15.1 Estimation: Coin Weight

How much does a nickel weigh?

1. Record an estimate that is:

| Too Low | About Right | Too High |
|---------|-------------|----------|
|         |             |          |

2. Explain your reasoning.

# Activity

## 15.2 Row Game: Equations

Work independently on your column. Partner A completes the questions in column A only and partner B completes the questions in column B only. Your answers in each row should match. Work on one row at a time and check if your answer matches your partner's before moving on. If you don't get the same answer, work together to find any mistakes.

Solve each equation.

| Row | Column A | Column B |
|-----|----------|----------|
| 1 | $3x - 1 = 5$ | $6x - 2 = 10$ |
| 2 | $4(x + 1) = 3x - 12$ | $4x + 4 = 3(x - 4)$ |
| 3 | $14x + 10 = 4x + 6$ | $7x + 5 = 2x + 3$ |
| 4 | $6x + 3 = 33$ | $4x + 2 = 22$ |
| 5 | $4x + 5y = 2 - 4x + 5y$ | $4x + 9y = 2 + 9y - 4x$ |
| 6 | $2x + 6y = 10$ | $5x + 15y = 25$ |

NAME _____ DATE _____ PERIOD _____

## Activity

### 15.3 What Were They Thinking?

Read each student's reasoning and answer the questions.

Jada says, "I know 4 nickels and 7 dollar coins weigh 76.7 grams. I know 4 nickels and 5 dollar coins weigh 60.5 grams. Here's what else I can figure out based on that:

- 2 dollar coins weigh 16.2 grams.

- 1 dollar coin weighs 8.1 grams.

- 5 dollar coins weigh 40.5 grams.

- 4 nickels weigh 20 grams.

- 1 nickel weighs 5 grams."

1. How did Jada figure out that 2 dollar coins weigh 16.2 grams?

2. Why might Jada have done that step first?

3. After Jada figured out how much 1 dollar coin weighed, why did she calculate how much 5 dollar coins weighed?

4. Once Jada knew how much 5 dollar coins weighed, how did she figure out how much 4 nickels weighed?

Priya says, "I know 9 plastic bricks and 3 number cubes weigh 39 grams, and 7 plastic bricks and 6 number cubes weigh 50.5 grams. Here's what I can figure out based on that:

- The weight of 18 plastic bricks and 6 number cubes is 78 grams.
- The weight of 11 plastic bricks is 27.5 grams.
- The weight of 1 plastic brick is 2.5 grams.
- The weight of 9 plastic bricks is 22.5 grams.
- The weight of 3 number cubes is 16.5 grams.
- The weight of 1 number cube is 5.5 grams."

1. Why is Priya's second step only about plastic bricks, not number cubes?

2. How does it help Priya to have a statement that's just about plastic bricks, and not number cubes?

3. Why might Priya have started by finding the weight of 18 plastic bricks and 6 number cubes?

4. After Priya figured out how much 1 plastic brick weighed, why did she calculate how much 9 plastic bricks weighed?

Lesson 2-16

# Elimination

NAME _____ DATE _____ PERIOD _____

**Learning Goal** Let's learn how to check our thinking when using elimination to solve systems of equations.

 ## Warm Up

### 16.1 Which One Doesn't Belong: Systems of Equations

Which one doesn't belong?

A.
$$\begin{cases} 3x + 2y = 49 \\ 3x + 1y = 44 \end{cases}$$

B.
$$\begin{cases} 3y - 4x = 19 \\ -3y + 8x = 1 \end{cases}$$

C.
$$\begin{cases} 4y - 2x = 42 \\ -5y + 3x = -9 \end{cases}$$

D.
$$\begin{cases} y = x + 8 \\ 3x + 2y = 18 \end{cases}$$

## Activity
### 16.2 Examining Equation Pairs

Here are some equations in pairs. For each equation:

- Find the $x$-intercept and $y$-intercept of a graph of the equation.
- Find the slope of a graph of the equation.

1. $x + y = 6$ and $2x + 2y = 12$

2. $3y - 15x = -33$ and $y - 5x = -11$

3. $5x + 20y = 100$ and $4x + 16y = 80$

4. $3x - 2y = 10$ and $4y - 6x = -20$

5. What do you notice about the pairs of equations?

6. Choose one pair of equations and rewrite them into slope-intercept form ($y = mx + b$). What do you notice about the equations in this form?

## Activity
### 16.3 Making the Coefficient

For each question,

- What number did you multiply the equation by to get the target coefficient?
- What is the new equation after the original has been multiplied by that value?

1. Multiply the equation $3x + 4y = 8$ so that the coefficient of $x$ is 9.

2. Multiply the equation $8x + 4y = -16$ so that the coefficient of $y$ is 1.

3. Multiply the equation $5x - 7y = 11$ so that the coefficient of $x$ is -5.

4. Multiply the equation $10x - 4y = 17$ so that the coefficient of $y$ is -8.

5. Multiply the equation $2x + 3y = 12$ so that the coefficient of $x$ is 3.

6. Multiply the equation $3x - 6y = 14$ so that the coefficient of $y$ is 3.

**Lesson 2-17**

# Number of Solutions in One-Variable Equations

NAME _____ DATE _____ PERIOD _____

**Learning Goal** Let's look at the number of solutions an equation may have.

## Warm Up
### 17.1 Notice and Wonder: Three Graphs

What do you notice? What do you wonder?

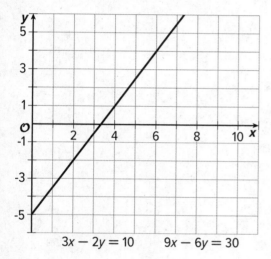

$$3x - 2y = 10 \qquad 9x - 6y = 30$$

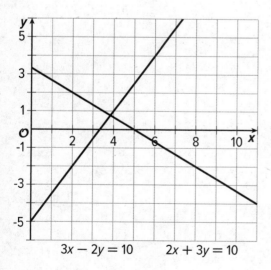

$$3x - 2y = 10 \qquad 2x + 3y = 10$$

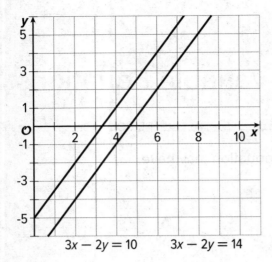

$3x - 2y = 10$     $3x - 2y = 14$

NAME _____ DATE _____ PERIOD _____

 ## Activity
### 17.2 How Many Answers?

How many values of $x$ make each equation true?

1. $3x + 1 = 10$

2. $2x + 12 = 2x + 10 + 2$

3. $2x = x + 2$

4. $3(x + 4) = 3x + 4$

5. $\dfrac{2x + 6}{2} = x + 6$

6. $0 = 0$

7. $x + 3x - 4 = 7\left(x - \dfrac{4}{7}\right)$

8. $0 = 6$

With your partner, discuss what you notice about the equations based on the number of solutions they have.

# Activity

**17.3 Write, Trade, Check**

1. Write an equation that has either 1, 0, or infinite solutions.

2. Trade your equation with your partner. Solve the equation you are given and determine the number of solutions.

3. Take turns explaining your reasoning with your partner.

4. Repeat the process with a new equation.

**Lesson 2-18**

# Inequalities in Context

NAME _____ DATE _____ PERIOD _____

**Learning Goal** Let's explore situations that compare values.

## Warm Up
### 18.1 Inequalities

Place a < or > to correctly complete the inequality.

**1.** 5 _____ 10

**2.** 5 _____ -10

**3.** -5 _____ -10

**4.** $\dfrac{1}{5}$ _____ $\dfrac{1}{10}$

**5.** $\dfrac{-1}{5}$ _____ $\dfrac{-1}{10}$

Elena and Lin are planning some cold-weather camping and are studying jackets and sleeping bags. Here is what they learned about some different equipment.

- A down jacket is rated as comfortable when the temperature is between -20°F and 20°F.

- A down sleeping bag is rated as comfortable when the temperature is -20°F or above.

- A synthetic sleeping bag is rated as comfortable when the temperature is 20°F or above.

1. What are 2 examples of temperatures that are comfortable for the down jacket?

2. For which gear would a temperature of -16.5°F be comfortable?

3. Is it possible to list all the temperatures for which the down jacket is comfortable? Explain your reasoning.

4. Which item's comfort rating matches each inequality?

   a. $x \geq 20$

   b. $-20 \leq x$

   c. $-20 \leq x$ and $x \leq 20$

NAME _____ DATE _____ PERIOD _____

**5.** Here are five graphs of inequalities. Match each graph to a situation in this activity.

A
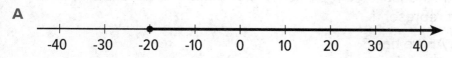

B
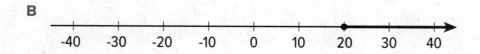

C
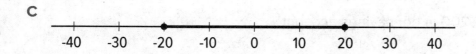

D
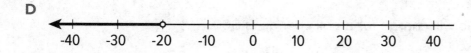

E
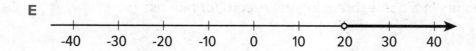

**6.** Are there any temperatures at which the sleeping bags and the jacket would all be comfortable?

**7.** Write one or more inequalities representing the range of temperatures in your area in the winter. Which gear, if any, would you recommend based on that?

## Activity

### 18.3 Representing Inequalities

For each statement, write an inequality to represent it. If a variable is used, be prepared to explain what it represents.

1. Han has 5 pencils, and Andre has 8.

2. Noah has more books than Kiran.

3. Clare has more than $200 in her savings account.

4. The most the mechanic will charge for an oil change is $60.

5. Diego scored 1,200 points in a game, breaking the record for highest score.

6. Jada is younger than Tyler.

7. Animal World has at least 400 different species of animals.

8. Mai's bowling score is more than Clare's and Han's combined.

**Lesson 2-19**

# Queuing on the Number Line

NAME _____ DATE _____ PERIOD _____

**Learning Goal** Let's use number line to reason about inequalities

 **Warm Up**

**19.1 Notice and Wonder: Shaded Number Line**

What do you notice? What do you wonder?

$4 > x$

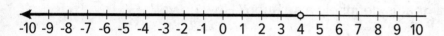

-10 -9 -8 -7 -6 -5 -4 -3 -2 -1 0 1 2 3 4 5 6 7 8 9 10

## Activity

### 19.2 Pick a Number

For each expression, pick a number you would like to evaluate, and tell whether it makes the inequality true. Be prepared to explain what made you choose your number.

1. $\frac{4}{3}y + 10 > 19$

   a. Pick a number you would like to test in place of $y$: -1, 0, 1, 3, 4, or 5. Explain why you chose your number.

   b. Does your number make the inequality true?

   c. What is a different number that is definitely a solution? How do you know?

   d. What is a different number that is definitely not a solution? How do you know?

2. $2.954x - 14.287 < 13.89$

   a. Pick a number you would like to test in place of $x$: -1, -0.5, 0, 0.5, 1, 3, 10, or 1,000. Explain why you chose your number.

   b. Does your number make the inequality true?

   c. What is a different number that is definitely a solution? How do you know?

   d. What is a different number that is definitely not a solution? How do you know?

**3.** $10 - 3y < 5$

    **a.** Pick a number you would like to test in place of $y$: -100, -3, -1, 0, $\frac{1}{3}$, $\frac{5}{3}$, 33, or 100. Explain why you chose your number.

    **b.** Does your number make the inequality true?

    **c.** What is a different number that is definitely a solution? How do you know?

    **d.** What is a different number that is definitely not a solution? How do you know?

**4.** $\frac{10x}{4} > \frac{3x}{5}$

    **a.** Pick a number you would like to test in place of $x$: -10, -5, -4, 0, 4, 5, 10, or 20. Explain why you chose your number.

    **b.** Does your number make the inequality true?

    **c.** What is a different number that is definitely a solution? How do you know?

    **d.** What is a different number that is definitely not a solution? How do you know?

## Activity

### 19.3 Matching Words and Symbols

For each inequality, write 3 values that make the inequality true, write 3 values that make it false, and choose a verbal description that matches the inequality.

**1.** $x > 13.5$

    **a.** Three values that make it true:

    **b.** Three values that make it false:

    **c.** Which verbal description best matches the inequality?

        **i.** $x$ is less than 13.5

        **ii.** $x$ is greater than 13.5

        **iii.** 13.5 is greater than $x$

**2.** $-27 < x$

    **a.** Three values that make it true:

    **b.** Three values that make it false:

    **c.** Which verbal description best matches the inequality?

        **i.** $x$ is less than -27

        **ii.** $x$ is greater than -27

        **iii.** -27 is greater than $x$

**3.** $x \geq \frac{1}{2}$ and $x \leq 2.75$

   **a.** Three values that make it true:

   **b.** Three values that make it false:

   **c.** Which verbal description best matches the inequality?

     **i.** $x$ is between $\frac{1}{2}$ and 2.75

     **ii.** 2.75 is less than $x$ is less than $\frac{1}{2}$

     **iii.** $x$ is greater than $\frac{1}{2}$

**4.** $x \geq -\frac{19}{4}$ and $x \leq \frac{1}{2}$

   **a.** Three values that make it true:

   **b.** Three values that make it false:

   **c.** Which verbal description best matches the inequality?

      **i.** $x$ is between $\frac{1}{2}$ and $-\frac{19}{4}$

      **ii.** $x$ less than $-\frac{19}{4}$

      **iii.** $x$ is between $-\frac{19}{4}$ and $\frac{1}{2}$

Lesson 2-20

# Interpreting Inequalities

NAME _____ DATE _____ PERIOD _____

**Learning Goal** Let's interpret the meaning of situations with phrases like "at least," "at most," and "up to."

 ## Warm Up
### 20.1 Math Talk: Solving Inequalities

Mentally solve for $x$.

$5x < 10$

$10 > 6x - 2$

$9x < 5 - 23$

$11(x - 3) < 46 - 2$

# Activity

## 20.2 Checking and Graphing Inequalities

Solve each inequality. Then, check your answer using a value that makes your solution true.

1. $-2x < 4$

   a. Solve the inequality.

   b. Check your answer using a value that makes your solution true.

2. $3x + 5 > 6x - 4$

   a. Solve the inequality.

   b. Check your answer using a value that makes your solution true.

3. $-3(x + 1) \geq 13$

   a. Solve the inequality.

   b. Check your answer using a value that makes your solution true.

NAME _____ DATE _____ PERIOD _____

For each statement:

a. Use a number line to show which values satisfy the inequality.

b. Express the statement symbolically with an inequality.

1. The elevator can lift up to 1,200 pounds. Let *x* represent the weight being lifted by the elevator.

+---+---+---+---+---+---+---+---+---+---+---+---+---+

2. Over the course of the senator's term, her approval rating was always around 53% ranging 3% above or below that value. Let *x* represent the senator's approval rating.

+---+---+---+---+---+---+---+---+---+---+---+---+---+---+---+---+---+---+---+

3. There's a minimum of 3 years of experience required. Let *x* represent the years of experience a candidate has.

+---+---+---+---+---+---+---+---+---+---+---+

## Activity

### 20.3 Card Sort: What's the Situation?

Your teacher will give you a set of cards that show a graph, an inequality, or a situation. Sort the cards into groups of your choosing. Be prepared to explain the meaning of your categories. Then, sort the cards into groups in a different way. Be prepared to explain the meaning of your new categories.

Lesson 2-21

# From One- to Two-Variable Inequalities

NAME _____ DATE _____ PERIOD _____

**Learning Goal** Let's look at inequalities in two dimensions.

## Warm Up
### 21.1 Describing Regions of the Plane

For each graph, what do all the ordered pairs in the shaded region have in common?

A

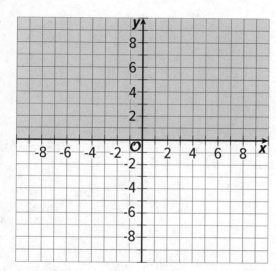

B

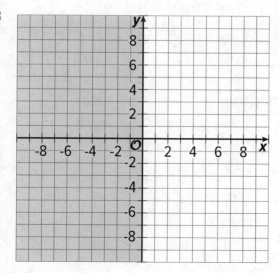

**C**

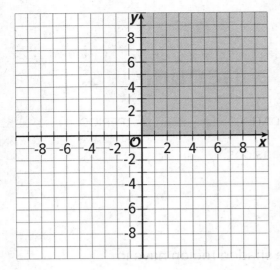

**D**

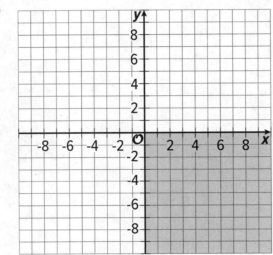

 ## Activity

### 21.2 More or Less

1. Write at least 3 values for $x$ that make the inequality true.

   a. $x < -2$

   b. $x + 2 > 4$

   c. $2x - 1 \leq 7$

NAME _____ DATE _____ PERIOD _____

**2.** Graph the solution to each inequality on a number line.

a.

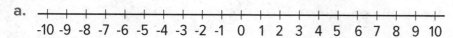

-10 -9 -8 -7 -6 -5 -4 -3 -2 -1 0 1 2 3 4 5 6 7 8 9 10

b.

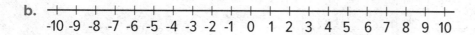

-10 -9 -8 -7 -6 -5 -4 -3 -2 -1 0 1 2 3 4 5 6 7 8 9 10

c.

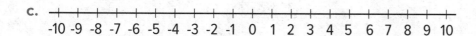

-10 -9 -8 -7 -6 -5 -4 -3 -2 -1 0 1 2 3 4 5 6 7 8 9 10

**3.** Using the inequality $x < -2$, write 3 coordinate pairs for which the $x$-coordinate makes the inequality true. Use the coordinate plane to plot your 3 points.

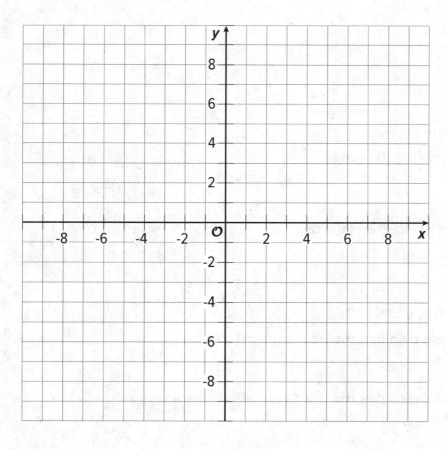

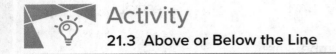

# Activity

## 21.3 Above or Below the Line

1. Graph the line that represents the equation $y = 3x - 4$

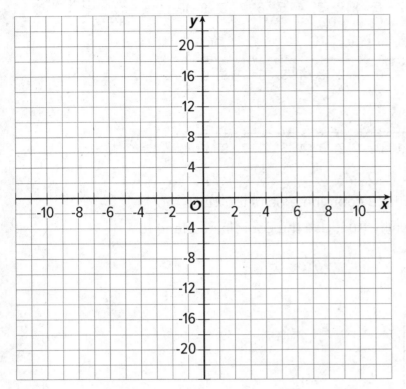

2. Is the point (4, 8) on the line?

   a. Explain how you know using the graph.

   b. Explain how you know using the equation.

3. Use the 3 points $(5, a), (-7, b)$ and $(c, 20)$

   a. Write values for $a, b,$ and $c$ so that the points are on the line.

   b. Write values for $a, b,$ and $c$ so that the points are above the line.

   c. Write values for $a, b,$ and $c$ so that the points are below the line.

**Lesson 2-22**

# Situations with Constraints

NAME _____ DATE _____ PERIOD _____

**Learning Goal** Let's study situations that have constraints.

## Warm Up
### 22.1 Graph Features of Inequalities

For each inequality:

1. What is the *x*-intercept of the graph of its boundary line?

2. What is the *y*-intercept of the graph of its boundary line?

3. Plot both intercepts, and then use a ruler to graph the boundary of the inequality.

$2y \geq 4x - 8$

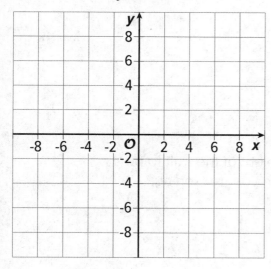

$2x + 3y < 12$

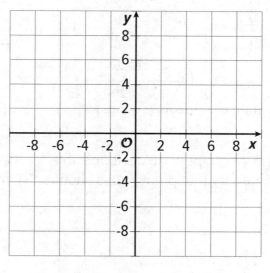

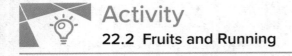

Write an equation that helps to answer the question about the situation. Then draw a graph that represents the equation.

1. Jada goes to an orchard to pick plums and apricots to make jam. She picks 20 pounds of fruit altogether. If she picks $a$ pounds of apricots, how many pounds of plums does she pick?

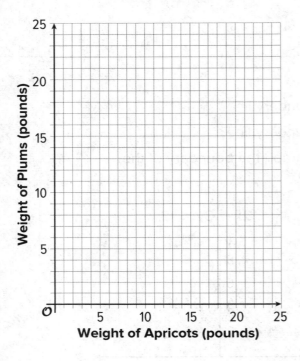

a. Consider the point (5, 16). Is it possible for the weight of the fruit to be represented by that point in this situation? Explain your reasoning.

NAME _____ DATE _____ PERIOD _____

**2.** In a video game, a character can run at a top speed of 30 miles per hour, but each additional pound that the character carries lowers the maximum running speed by 1 mile per hour. What is the maximum running speed of the character when they are carrying *w* pounds?

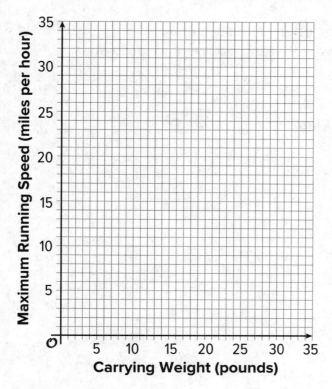

**a.** Consider the point (10, 15). Is it possible for a character in this game to be represented by that point in this situation? Explain your reasoning.

## Activity

### 22.3 Matching Graphs and Inequalities

1. Take turns with your partner to match graphs, inequalities, and constraints.

   a. For each match that you find, explain to your partner how you know it's a match.

   b. For each match that your partner finds, listen carefully to their explanation. If you disagree, discuss your thinking and work to reach an agreement.

**Lesson 2-23**

# Modeling Constraints

NAME _____ DATE _____ PERIOD _____

**Learning Goal** Let's represent some situations from banking and insurance.

## Warm Up

### 23.1 Notice and Wonder: The Wonderful World of Finance

What do you notice? What do you wonder?

1. Jada received $100 on her birthday. She has a savings account and a checking account that she can deposit the money in.

**2.** Han's uncle is an insurance agent. He sells customers two types of car insurance policies: a cheap one and an expensive one. The cheap car insurance has a value of $7,000 and the expensive one has a value of $18,000. His goal for the month is to sell policies valuing over $400,000 total.

## Activity

### 23.2 Insurance Policies

Han's uncle is an insurance agent. He sells customers two types of car insurance policies: a cheap one and an expensive one. The cheap car insurance has a value of $7,000 and the expensive one has a value of $18,000. His goal for the month is to sell policies valuing over $400,000 total.

**1.** List some different amounts of each policy Han's uncle could sell.

**2.** What calculations could you do to check whether Han's uncle reached his goal?

**3.** What could you compare your answers to in order to see if he reached the goal?

NAME _____ DATE _____ PERIOD _____

**4.** Complete the table using the values from the previous questions.

| Number of Cheap Policies Sold | Number of Expensive Policies Sold | Calculation | Check |
|---|---|---|---|
|  |  |  |  |
|  |  |  |  |
|  |  |  |  |
| $x$ | $y$ |  |  |

**5.** Write an inequality using number of cheap policies, $x$, and number of expensive policies, $y$. The inequality should be true if Han's uncle meets his goal.

## Activity

### 23.3 Row Game: Writing Inequalities from Situations

Your teacher will assign you a set. Work only on the problems in your set. Work on one question at a time and check whether your answer matches your partner's before moving on.

Set A

**1.** Clare has $25.00 to spend on souvenirs during her class trip to Washington, D.C. She wants to buy some souvenirs from the Air & Space Museum and some from the National Museum of African American History and Culture. She might not spend all of her money. Let $x$ represent the amount of money she spends at Air & Space and $y$ represent the amount of money she spends at the African American museum.

   **a.** What is one ordered pair $(x, y)$ that will work in this situation?

**b.** Write an inequality in terms of $x$ and $y$ that shows what Clare can spend on souvenirs.

2. Dried apricots have 10 grams of sugar per ounce. Cashews have 2 grams of sugar per ounce. Diego wants to make bags of trail mix with no more than 50 grams of sugar per bag. Let $x$ represent the number of ounces of apricots in a bag and $y$ represent the number of ounces of cashews in each bag.

   **a.** What is one ordered pair $(x, y)$ that will work in this situation?

   **b.** Write an inequality in terms of $x$ and $y$ that shows how many ounces of dried apricots and cashews Diego can include in his trail mix bags.

3. The band is raising money for their trip to Orlando. Each student needs to raise at least $250. They are selling candles which earn $7 each, and poinsettias which earn $15 each. Let $x$ represent the number of candles sold and $y$ represent the number of poinsettias sold.

   **a.** What is one ordered pair $(x, y)$ that will work in this situation?

   **b.** Write an inequality in terms of $x$ and $y$ that shows how many candles and poinsettias each student needs to sell.

4. Mai is trying to earn at least $75 toward prom-related expenses. Her mom has offered to pay her $3.00 every time she cleans the cat litter, and $5.00 every time she walks the dog. Let $x$ represent the number of times she cleans the cat litter and $y$ represent the number of times she walks the dog.

   **a.** What is one ordered pair $(x, y)$ that will work in this situation?

   **b.** Write an inequality in terms of $x$ and $y$ that shows how many times Mai could walk the dog and clean the cat litter to meet her goal.

NAME _____ DATE _____ PERIOD _____

## Set B

1. Lin's library sets a maximum of 25 items that can be checked out at one time. Lin likes to check out books and DVDs. Let $x$ represent the number of books Lin checks out, and $y$ represent the number of DVDs Lin checks out.

   **a.** What is one ordered pair $(x, y)$ that will work in this situation?

   **b.** Write an inequality in terms of $x$ and $y$ that shows how many books and DVDs Lin can check out.

2. Noah is sending a care package to his cousin in the military. He has saved $50 to spend. His cousin's favorite items are movies, which Noah found on sale for $10 each, and energy bars, which are $2 each. Let $x$ represent the number of movies Noah buys, and $y$ represent the number of energy bars. Noah doesn't have to spend all of the money on this care package.

   **a.** What is one ordered pair $(x, y)$ that will work in this situation?

   **b.** Write an inequality in terms of $x$ and $y$ that shows how many movies and energy bars Noah can send his cousin.

3. A group of teachers is ordering school supplies online. They need pencils, which are $7 a box, and paper, which is $15 a box. They get free shipping on orders of $250 or more. Let $x$ represent the number of boxes of pencils they buy, and $y$ represent the number of boxes of paper they buy.

   **a.** What is one ordered pair $(x, y)$ that will work in this situation?

   **b.** Write an inequality in terms of $x$ and $y$ that shows how many boxes of pencils and paper the teachers could buy to get free shipping.

**4.** Priya is helping her cousins at their farm stand. Her aunt has asked them to try to sell at least 75 pounds of tomatoes by noon. They sell tomatoes in 3-pound and 5-pound bags. Let $x$ represent the number of 3-pound bags of tomatoes they sell and $y$ represent the number of 5-pound bags they sell.

    **a.** What is one ordered pair $(x, y)$ that will work in this situation?

    **b.** Write an inequality in terms of $x$ and $y$ that shows how many 3-pound bags and 5-pound bags Priya could sell to meet her goal.

Lesson 2-24

# Reasoning with Graphs of Inequalities

NAME _____ DATE _____ PERIOD _____

**Learning Goal** Let's solve riddles about inequalities.

## Warm Up
### 24.1 Notice and Wonder: Shady Graphs

What do you notice? What do you wonder?

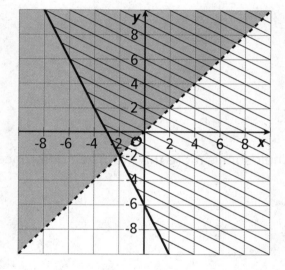

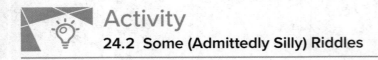

# Activity

### 24.2 Some (Admittedly Silly) Riddles

Each riddle consists of two conditions. For each riddle,

- find a pair of numbers that satisfy the first condition.
- find a pair of numbers that satisfy the second condition.
- find a pair of numbers that satisfy both conditions.
- determine whether there could be more than one solution.
- write a system of equations or inequalities that represent the riddle.

1. I'm thinking of two numbers. Their sum is 15. Their difference is 1.

2. I'm thinking of two numbers. Their sum is more than 15. Their difference is more than 1.

3. I'm thinking of two numbers. One is more than twice the other. Their sum is less than 30.

**4.** Think of your own riddles about a pair of numbers.

    **a.** Make a riddle with multiple pairs of numbers that will work.

    **b.** Make a riddle with one pair of numbers that will work.

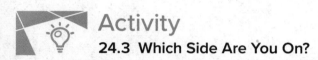

## Activity

### 24.3 Which Side Are You On?

For each question:

• Draw a graph that represents the equation or equations.

• Determine how many regions the lines split the plane into.

• Play a game with your partner

    • Partner A: Think of one of the regions.

    • Partner B: Without pointing to the coordinate plane, use the lines to ask questions of your partner about the region they have chosen. When you think you know which region they are thinking of, point to the region and ask your partner if you are correct. If you are correct, shade the entire region.

    • Change roles for the next question.

**1.** $x = 3$

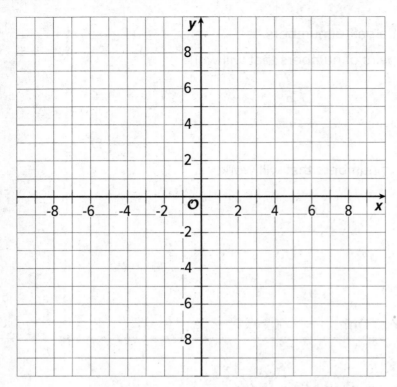

**2.** $y = -1$ and $x = 2$

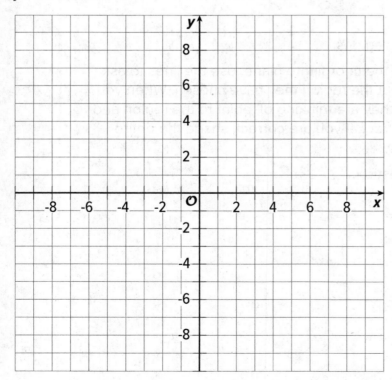

NAME _____ DATE _____ PERIOD _____

**3.** $y = x$ and $y = -x$

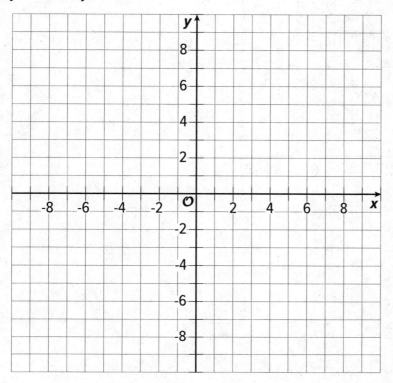

**4.** $y = 2x + 2$ and $y = 2x - 3$

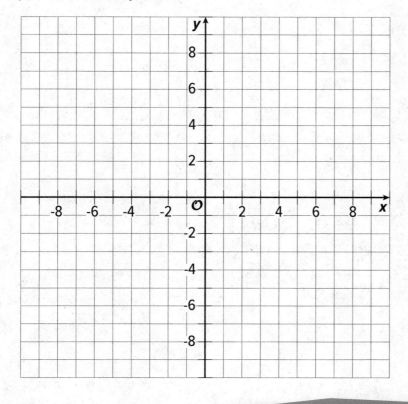

**5.** $x + y = 4$ and $x - y = 6$

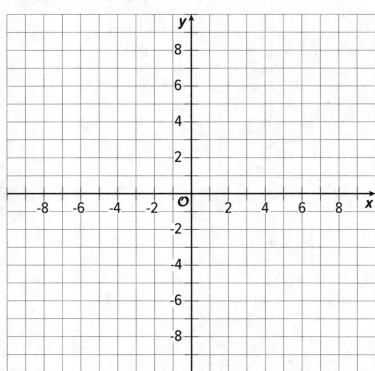

Lesson 2-25

# Representing Systems of Inequalities

NAME _____ DATE _____ PERIOD _____

**Learning Goal** Let's find and represent solutions to situations involving inequalities.

## Warm Up
### 25.1 Which One Doesn't Belong: Splash Zone!

Which one doesn't belong?

**A:** Clare's family wants to

- spend at least 4 hours at the amusement park

- spend more time in the Splash Zone than riding rides

**B:** Jada's family wants to

- be at the amusement park from 4 p.m. to 8 p.m.
- spend most of their time riding rides

**C:** Priya's family wants to

- spend 2 hours at Splash Zone
- 2 hours riding rides

**D:** Diego's family wants to

- spend no more than 6 hours at the amusement park
- spend at least twice as long riding rides as they spend at Splash Zone

NAME _____ DATE _____ PERIOD _____

# Activity
## 25.2  Amusing Solutions

For each family, let *x* be the amount of time each family spends riding rides, and *y* be the amount of time each family spends at the Splash Zone.

List one or more ordered pairs (*x*, *y*) that would fit the constraints. If you can only list one, explain why you can only list one.

**1.** Clare's family wants to spend at least 4 hours at the amusement park, and they want to spend more time in the Splash Zone than riding rides.

**2.** Jada's family wants to be at the amusement park from 4 p.m. to 8 p.m., and they want to spend most of their time riding rides.

**3.** Priya's family wants to spend 2 hours at Splash Zone and 2 hours riding rides.

**4.** Diego's family wants to spend no more than 6 hours at the amusement park, and they want to spend at least twice as long riding rides as they spend at Splash Zone.

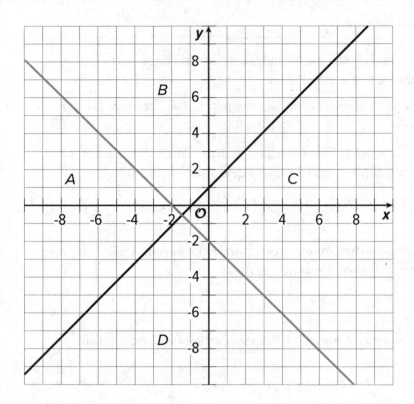

1. The graph shows the lines $y = x + 1$ and $y = -x - 2$. Which line represents $y = x + 1$?

2. For each of the 4 regions, write a coordinate pair for a point in that region.

3. Change the equations represented by the lines into inequalities so that the region labeled as A is shaded by both inequalities.

   a. $y$ _____ $x + 1$

   b. $y$ _____ $-x - 2$

4. Use the coordinate pairs you chose for region A to check your inequalities.

**Lesson 2-26**

# Testing Points to Solve Inequalities

NAME _____ DATE _____ PERIOD _____

**Learning Goal** Let's critique some peoples' reasoning.

 Warm Up

**26.1 Math Talk: Solving Equations**

Solve each equation mentally.

$3x + 5 = 14$

$3(x - 1) + 5 = 14$

$3x - 3 + 5 = 14$

$3(1 - x) + 5 = 14$

Andre is working on $\frac{5x}{3} - 1 < \frac{2}{3}$. He figured out that when $x = 1$, $\frac{5(1)}{3} - 1 = \frac{2}{3}$. He tested all these points:

- When $x = -1$, $\frac{5(-1)}{3} - 1 = \frac{-8}{3}$, $\frac{-8}{3} < \frac{2}{3}$
- When $x = 0$, $\frac{5(0)}{3} - 1 = -1$, $-1 < \frac{2}{3}$
- When $x = 2$, $\frac{5(2)}{3} - 1 = \frac{7}{3}$, $\frac{7}{3} > \frac{2}{3}$
- When $x = 3$, $\frac{5(3)}{3} - 1 = 4$, $4 > \frac{2}{3}$

Based on these results, Andre determines that solutions for $x$ should include -1 and 0, but not 2 or 3.

1. Andre is frustrated with how much computation he had to do. What advice would you give him about how many numbers to test and which ones to test?

2. Mai was trying to solve $10 - 3x > 7$. She saw that when $10 - 3(1) = 7$. She reasoned, "Because the problem has a greater than sign, I wrote $x > 1$." Mai skipped the step of testing points, and that led to an error.

   a. Help Mai test points to determine the correct solution to the inequality.

   b. Explain to Mai what went wrong with her reasoning.

NAME _____ DATE _____ PERIOD _____

## Activity
### 26.3 Error!

Each of these solutions has something wrong. Circle the place that is wrong and write a correction.

**1.** $2x + 3 = 5x - 4$
$$5x = 5x - 4$$
$$0 = -4$$

**2.** $5x + 4 = 10 - 5x$
$$4 = 10$$

**3.** $2x + 8 = 2x + 100$
$$4x + 8 = 100$$
$$x + 2 = 50$$
$$x = 48$$

**4.** $5x + 50 = 20x$

$\qquad 50 = 25x$

$\qquad\ \ 2 = x$

**5.** $2(x + 8) = 16$

$\quad 2x + 16 = 16$

$\qquad\ \ 2x = 0$

No solution

**6.** $(x + 3) + 5 = 5$

$\qquad x + 3 = 0$

$\qquad\quad\ x = 3$

# Two-variable Statistics

Using data on bee populations, you can fit a model that can help predict future trends. You will learn more about fitting models to data in this unit.

## Topics
- Two-way Tables
- Scatter Plots
- Correlation Coefficients
- Estimating Lengths

# Two-variable Statistics

Lesson 3-1

# Human Frequency Table

NAME _____ DATE _____ PERIOD _____

**Learning Goal** Let's use tables to organize data.

## Warm Up

### 1.1 Estimation: Percents

What percentage of the graph is labeled B?

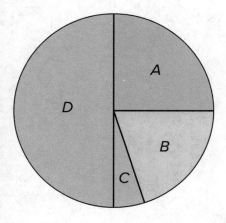

1. Record an estimate that is:

| Too Low | About Right | Too High |
|---------|-------------|----------|
|         |             |          |

2. Explain your reasoning.

# Activity

## 1.2 Human Frequency Table

Follow your teacher's instructions to create a two-way table with your class. Then, using the data from the table your class creates, answer the questions:

1.  How many students prefer tablets?

2.  How many students prefer basketball?

3.  How many students prefer both laptops and prefer baseball?

4.  Of the students who prefer tablets, how many prefer basketball?

NAME _____  DATE _____  PERIOD _____

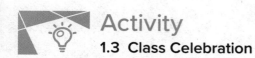

## Activity
### 1.3  Class Celebration

Han's teacher is planning a celebration for the class. Here is a table that displays the students' preferences for the celebration.

|                  | Prefers Monday | Prefers Friday | Total |
|------------------|----------------|----------------|-------|
| Prefers Indoors  | 4              | 8              |       |
| Prefers Outdoors | 6              | 9              |       |
| Total            |                |                |       |

Use the table to answer the questions.

1. Complete the table with the missing values.

2. How many students were surveyed?

3. How many students prefer Monday?

4. How many students prefer Friday?

5. On which day should Han's teacher plan to have the celebration?

6. How many students prefer to stay indoors?

7. How many students prefer to go outdoors?

8. How many students prefer Friday and prefer to stay indoors?

9. How many students prefer Monday and prefer to go outdoors?

**Lesson 3-2**

# Talking Percents

NAME _____ DATE _____ PERIOD _____

**Learning Goal** Let's explore percentages

 ## Warm Up
### 2.1 Math Talk: Percents

Evaluate mentally.

50% of 200

25% of 200

6% of 200

3.2% of 200

**2.2 Shapes Galore**

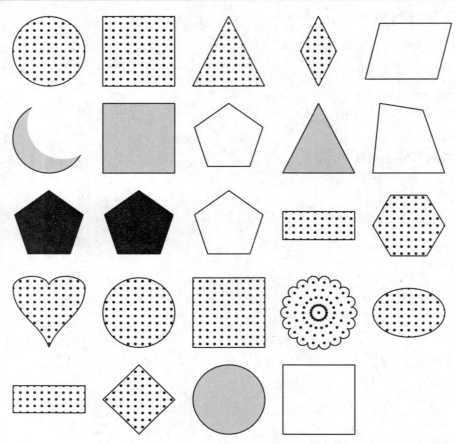

Use the image to fill in one word or number for each of the blanks. Round each percentage to the nearest whole number.

1. _____ of the 13 shapes with dots are circles. _____ % of the shapes with dots are circles.

2. 5 of the _____ _____ are white. _____% of the shapes are white.

3. _____ of the _____ shapes are quadrilaterals. _____% of the shapes are quadrilaterals.

4. _____ of the 4 _____ are _____.

   _____ % of the 4 _____ are _____.

5. a. _____ of the _____ _____ are _____.

   _____ % of the _____ are _____.

NAME _____ DATE _____ PERIOD _____

## Activity
### 2.3 Favorite Movies and Favorite Classes

For each question:

- Explain to a partner what categories you would use to calculate the percentage.

- Write an expression to use to find the percentage.

- Calculate the percentage to the nearest whole number.

| | Prefer Sci-Fi Movies | Prefer Drama Movies | Total |
|---|---|---|---|
| Prefer Science Class | | | |
| Prefer English Class | | | |
| Total | | | |

1. What percentage of the class prefer sci-fi movies?

2. What percentage of the class prefer science class?

3. What percentage of the class prefer dramas?

**4.** What percentage of the class prefer science class and dramas?

**5.** What percentage of the class prefer English class?

**6.** Of the students who prefer dramas, _____ % prefer English class.

**7.** Of the students who prefer science class, _____ % prefer sci-fi movies.

# Associations and Relative Frequency Tables

NAME _____ DATE _____ PERIOD _____

**Learning Goal** Let's explore relative frequency tables

## Warm Up
### 3.1 Estimation

What percentage of the graph is labeled C?

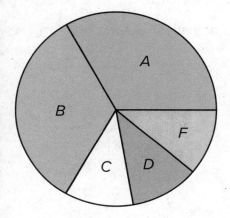

**1.** Record an estimate that is:

| Too Low | About Right | Too High |
|---------|-------------|----------|
|         |             |          |

**2.** Explain your reasoning.

# Activity

## 3.2 Relative Frequency Tables

The relative frequency tables display data collected from 230 students.

1.

| | Participates in Afterschool Activity | No Afterschool Activity | Total |
|---|---|---|---|
| **Arrives Home within 1 Hour of School Dismissal** | 3% | 40% | 43% |
| **Arrives Home 2 or More Hours After School Dismissal** | 42% | 15% | 57% |
| **Total** | 45% | 55% | 100% |

a. What percentage of students participate in after—school activities? How many students participate in after—school activities?

b. What percentage of students arrive home 2 or more hours after dismissal? How many students arrive home 2 or more hours after school dismissal?

NAME _____ DATE _____ PERIOD _____

2.

| | Aspiring Professional Athlete | Aspiring STEM Career | Total |
|---|---|---|---|
| Prefer Physical Education | 77% | 23% | 100% |
| Prefer Math | 18% | 82% | 100% |

a. What percentage of students who prefer math aspire to have a career in STEM?

b. What percentage of students who prefer physical education aspire to have a career in STEM?

c. Are these two percentages close?

d. Is there evidence of an association between students' career aspirations and subject preference? Explain your reasoning.

**3.**

|  | 9th Grade | 12th Grade |
|---|---|---|
| Curfew | 95% | 90% |
| No Curfew | 5% | 10% |
| Total | 100% | 100% |

a. Of the students in 12th grade, what percentage have a curfew?

b. Of the students in 9th grade, what percentage have a curfew?

c. Is there evidence of an association between students' grade level and whether they have a curfew? Explain your reasoning.

## Activity

### 3.3 Associate Your Variables

1. Invent a pair of variables that you think will have an association. Explain your reasoning.

2. Invent a pair of variables that you think will not have an association. Explain your reasoning.

Lesson 3-4

# Interpret This, Interpret That

NAME _____ DATE _____ PERIOD _____

**Learning Goal** Let's explore linear models

## Warm Up
### 4.1 Math Talk: Units

Mentally calculate each value.

5 granola bars cost $20. How much is 1 worth?

A car travels at a constant speed and goes 100 miles in 2.5 hours. How fast is the car travelling in miles per hour?

Tyler can do 50 sit-ups in 4 minutes. What is his average sit-ups per minute?

3 ounces of yeast flakes costs $4.29. What is the cost for 1 ounce?

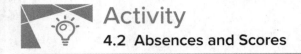

# Activity

## 4.2 Absences and Scores

Here are a table and scatter plot representing the number of students' absences and their final exam scores.

| Student | Number of Absences | Final Exam Score | Student | Number of Absences | Final Exam Score |
|---------|--------------------|-----------------|---------|--------------------|-----------------|
| A | 1 | 94 | M | 7 | 68 |
| B | 5 | 71 | N | 8 | 65 |
| C | 1 | 98 | O | 20 | 42 |
| D | 5 | 70 | P | 10 | 63 |
| E | 3 | 67 | Q | 11 | 63 |
| F | 2 | 94 | R | 20 | 50 |
| G | 6 | 71 | S | 15 | 67 |
| H | 4 | 89 | T | 16 | 40 |
| I | 5 | 77 | U | 4 | 86 |
| J | 0 | 90 | V | 8 | 82 |
| K | 2 | 91 | W | | |
| L | 11 | 60 | | | |

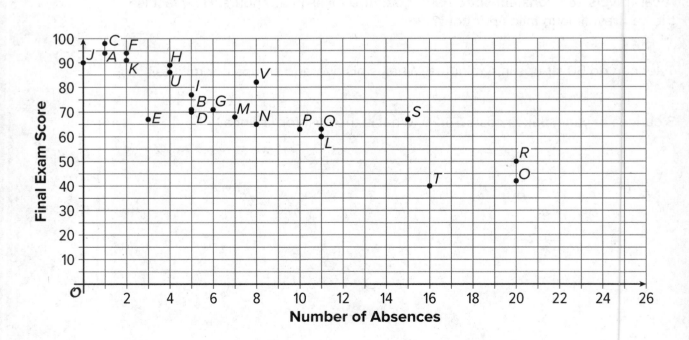

NAME _____ DATE _____ PERIOD _____

1. What are the coordinates of the point in the scatter plot that represents student *G*?

2. What are the coordinates of the point in the scatter plot that represents student *R*?

3. What is the final exam score of the student who has perfect attendance?

4. What are the final exam scores of the students with the most absences?

5. How many absences does the student with the highest score have?

6. How many absences does the student with the lowest score have?

7. If student *W* has 12 absences, what final exam score do you estimate the student will have? Plot this point on the scatter plot.

## Activity
### 4.3 Elevator Weights

Here is a linear model of the weight of an elevator and the number of people on the elevator.

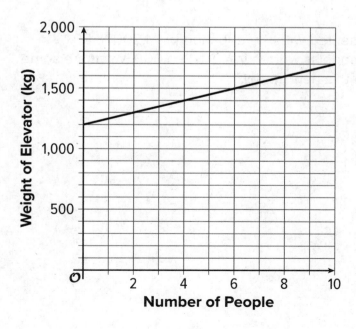

1. Find these values. Explain your reasoning.

   a. the weight of the elevator when 6 people are on it

   b. the number of people on the elevator when it weighs 1,400 kg

   c. the weight of the elevator when no people are on it

   d. the increase in elevator weight for each additional person according to the model

2. Which of your answers corresponds to the slope of the line in the graph?

3. Which of your answers corresponds to the $y$-intercept of the line in the graph?

4. This model can be represented with the equation $y = 1{,}200 + 50x$. An equation for a different model is written $y = 70x + 1{,}000$. What are some things you can say about this new model?

Lesson 3-5

# Goodness of Fit

NAME _____ DATE _____ PERIOD _____

**Learning Goal** Let's explore lines and their goodness of fit for data

## Warm Up
### 5.1 What's the Rate?

Each situation can be modeled using a linear equation. Describe the rate of change for each situation.

1. Andre started his no-interest savings account with $1,000. He makes the same deposit each week, and there is $1,600 in the account after 6 weeks.

2. Kiran starts with $748 in his checking account. After 4 weeks of spending the same amount each week, he has $716 left.

Here are 3 copies of the same scatter plot. Each student tries to draw a line that models the data well.

Noah says his line fits the data well because the line connects the leftmost point to the rightmost point.

Andre says his line fits the data well because it passes directly through as many points as possible.

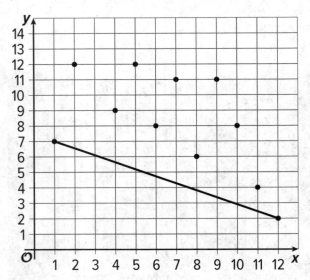

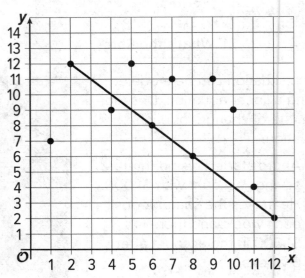

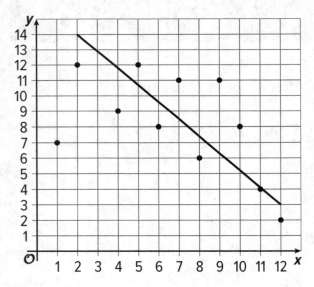

Lin says her line fits the data well because the points are somewhat evenly arranged around the line with about half the points above the line and half the points below the line.

Do you agree with any of these students? Explain your reasoning.

NAME _____ DATE _____ PERIOD _____

## Activity
### 5.3 What Fits?

**1.** Look at the scatter plots, and determine which one is best modeled by a linear model.

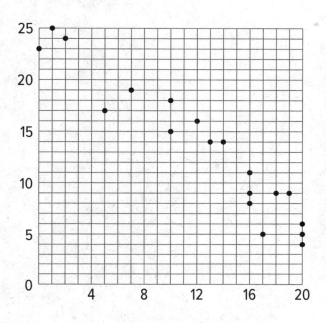

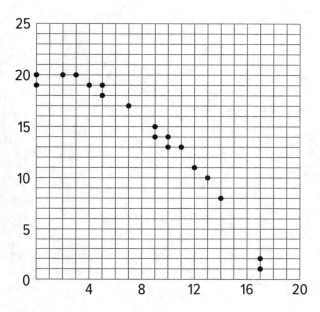

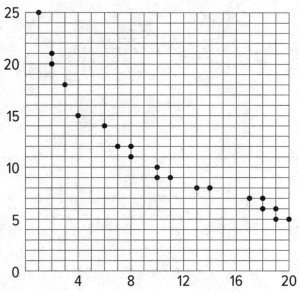

**2.** Draw a linear model that fits the data well on the appropriate scatter plot. Compare your line with a partner's. If your lines are different, determine which line is the better fit line.

Lesson 3-6

# Actual Data vs. Predicted Data

NAME _____ DATE _____ PERIOD _____

**Learning Goal** Let's explore linear models that are fit to data

 ## Warm Up
### 6.1 Which One Doesn't Belong: Data Representations

Which one doesn't belong?

**A**

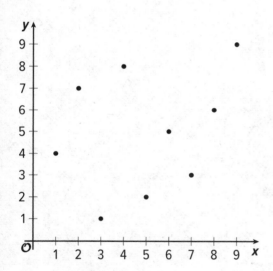

**B**

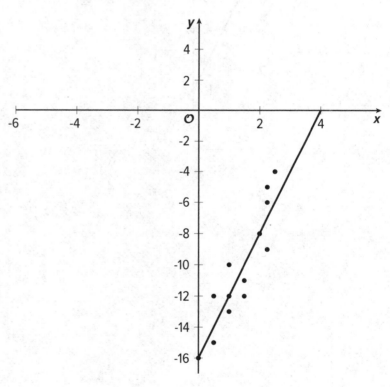

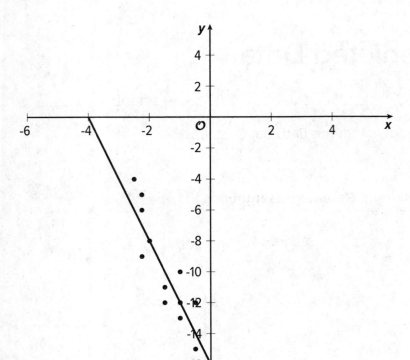

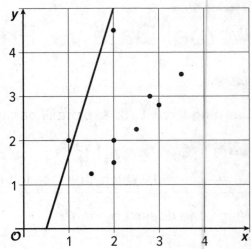

NAME _____ DATE _____ PERIOD _____

# Activity
## 6.2  Predicting Sales

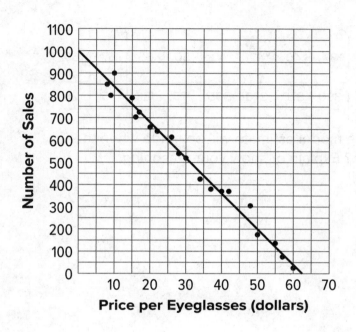

Here are a graph and a table showing the number of sales of eyeglasses based on the price in dollars. The model, represented by $y = 1{,}000 - 16x$, is graphed with a scatter plot. Use the graph and the table to answer the questions.

| Price Per Eyeglasses (dollars) | 8 | 9 | 10 | 15 | 16 | 17 | 20 | 22 | 26 | 28 |
|---|---|---|---|---|---|---|---|---|---|---|
| Number of Sales | 850 | 800 | 900 | 789 | 703 | 725 | 658 | 640 | 614 | 540 |

| Price Per Eyeglasses (dollars) | 30 | 34 | 37 | 40 | 42 | 48 | 50 | 55 | 57 | 60 |
|---|---|---|---|---|---|---|---|---|---|---|
| Number of Sales | 520 | 425 | 380 | 370 | 370 | 305 | 175 | 136 | 75 | 25 |

1. How many sales does the model estimate will be made when the eyeglasses are $50 each? Explain or show your reasoning.

2. How many sales were actually made when the eyeglasses were $50 each?

3. How many times did the model estimate fewer sales than what were actually made? List the coordinates.

4. How many times were the predicted number of sales and actual number of sales equivalent? List the coordinates.

5. Find a point for which the model predicted there would be at least 25 more sales than were actually made?

NAME _____ DATE _____ PERIOD _____

## Activity

### 6.3 Predictions

Priya's family keeps track of the number of miles on each trip they take over the summer and the amount spent on gas for the trip. The model, represented by $y = 50 + 0.15x$, is graphed with a scatter plot.

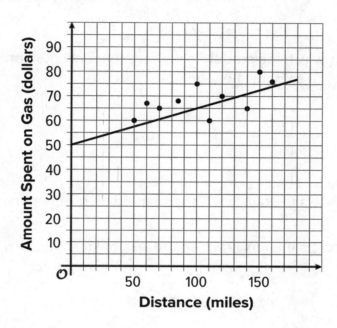

Use the graph and equation to complete the table. Then, use the graph, equation, and table to answer the questions.

| Distance (miles) | Amount Spent on Gas (dollars) | Estimated Amount Spent on Gas (dollars) |
|---|---|---|
| 50 | 60 | |
| 70 | 65 | |
| 100 | 75 | |
| 60 | 67 | |
| 110 | 60 | |
| 140 | 65 | |
| 80 | 68 | |
| 150 | 80 | |
| 160 | 76 | |

1. When Priya's family drove 85 miles, they spent $68 on gas. How much did they expect to spend based on the linear model?

2. How far had the family gone when they spent $80 on gas?

3. How far does the model estimate the family should have driven when they spent $80 on gas?

4. Are there any instances for which the model's estimated amount spent on gas is equivalent to the actual amount spent on gas?

5. Circle one option.

   - In general, the model predicts the family will spend more on gas than they actually spend.

   - In general, the model predicts the family will spend less on gas than they actually spend.

Lesson 3-7

# Confident Models

NAME _____ DATE _____ PERIOD _____

**Learning Goal** Let's explore our confidence in linear models

## Warm Up
### 7.1 Math Talk: Ordering Decimals

Mentally order the numbers from least to greatest.

20.2, 18.2, 19.2

-14.6, -16.7, -15.1

-0.43, -0.87, -0.66

0.50, -0.52, 0.05

1. Here are scatter plots that represent various situations. Order the scatter plots from "A linear model is not a good fit for the data" to "A linear model is an excellent fit for the data."

**A**

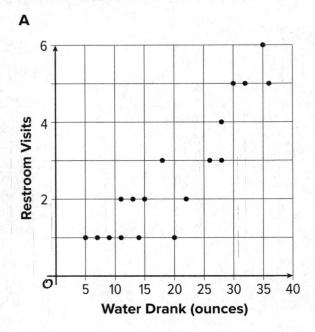

**B**

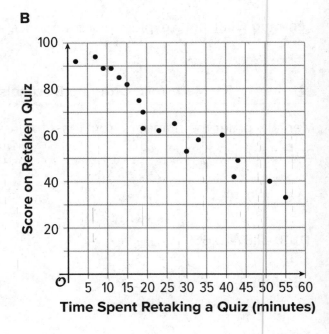

**C**

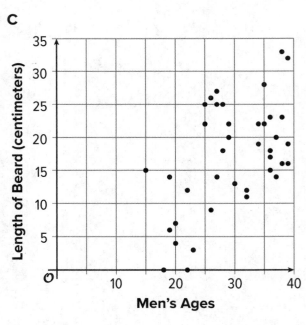

**D**

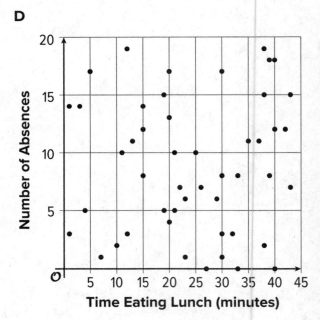

NAME _____ DATE _____ PERIOD _____

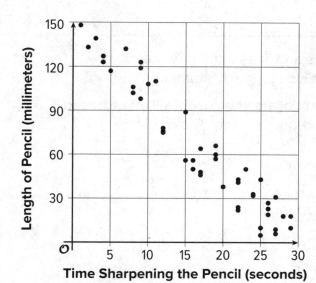

**E**

Length of Pencil (millimeters)

Time Sharpening the Pencil (seconds)

2. Here are two scatter plots including a linear model. For each model, determine the $y$ when $x$ is 15. Which model prediction do you think is closer to the real data? Explain your reasoning.

**Graph F.** $y = 150 - 5x$

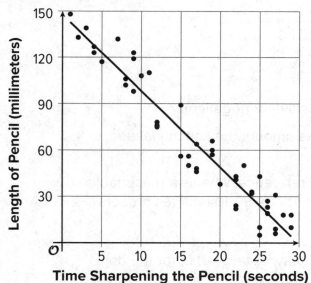

Length of Pencil (millimeters)

Time Sharpening the Pencil (seconds)

**Graph G.** $y = 100 - 1.3x$

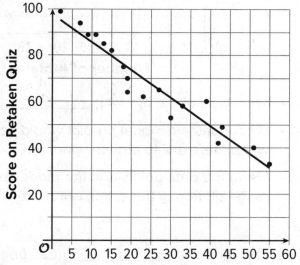

Score on Retaken Quiz

Time Spent Retaking a Quiz (minutes)

Here are situations represented with graphs and lines of fit. Use the information given to complete the missing fields for each situation.

1. The model predicts how much money, in dollars, the coach will make based on how many athletes sign up for one-on-one training. The model is represented with the equation $y = 200 + 25x$.

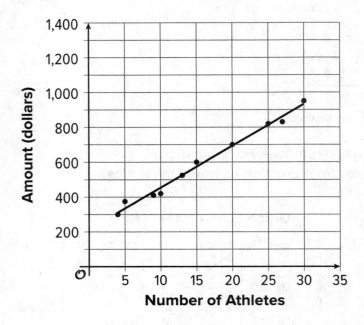

- The slope of the model is _____ (positive or negative).

- What does the model predict would be the amount the coach makes when there are 10 athletes present?

- Using the data points and the model as a reference, what is a reasonable range of money the coach will make when there are 10 athletes present?

- This model is a _____ (great, good, okay, or terrible) fit for the data.

- Using numbers between 0 and 1, rate your confidence in the model where 0 is no confidence and 1 is total confidence.

NAME _____ DATE _____ PERIOD _____

**2.** The model predicts the annual salary of a worker in a certain government position based on years of experience. The model is represented with the equation $y = 1.5x + 35$.

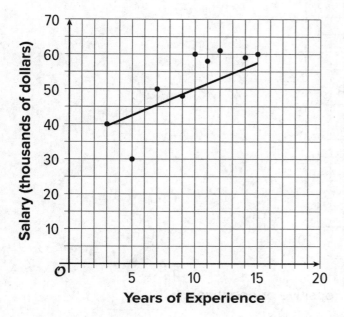

- The slope of the model is _____ (positive or negative).

- What does the model predict would be the employee's salary when the employee has 10 years of experience?

- Using the data points and the model as a reference, what is a reasonable range for the salary of a worker based on 10 years of experience?

- This model is a _____ (great, good, okay, or terrible) fit for the data.

- Using numbers between 0 and 1, rate your confidence in the model where 0 is no confidence and 1 is total confidence.

3. The model predicts the number of absences a school will have based on the number of incentives given per month. The model is represented with the equation $y = -2.18x + 54.78$.

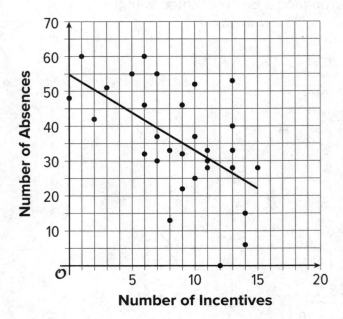

**Number of Incentives**

- The slope of the model is _____ (positive or negative).

- What does the model predict would be the number of absences when 10 incentives are given for the month?

- Using the data points and the model as a reference, what is a reasonable number of absences when there are 10 incentives given?

- This model is a _____ (great, good, okay, or terrible) fit for the data.

- Using numbers between 0 and 1, rate your confidence in the model where 0 is no confidence and 1 is total confidence.

Lesson 3-8

# Correlations

NAME _____ DATE _____ PERIOD _____

**Learning Goal** Let's explore correlations

## Warm Up

### 8.1 Notice and Wonder: Correlations

What do you notice? What do you wonder?

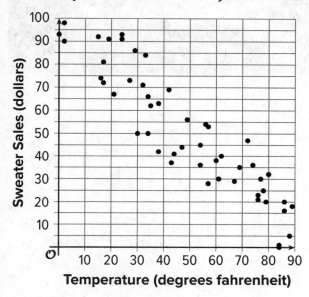

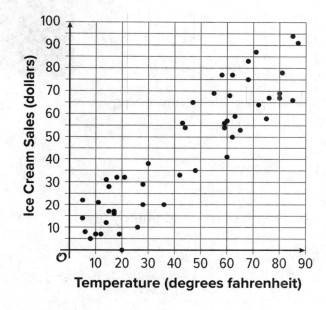

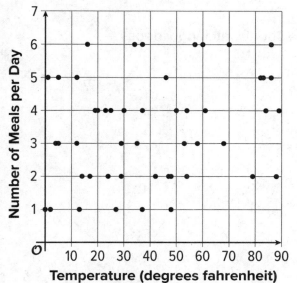

## Activity

### 8.2  Variable Relationships

1. For each pair of variables, do you expect there to be a relationship? That is, do you think a change in one variable is accompanied by a change in the other variable? How do you expect the second variable to change if the first variable is increased?

   a. hours of sleep and energy level

   b. length of hair and energy level

   c. number of school events each week and time spent watching videos online each week

   d. temperature and watermelon sales

NAME _____ DATE _____ PERIOD _____

**2.** Some data is collected for each pair of variables listed and represented by a scatter plot. For each pair of variables, how do the scatter plots support or contradict your answers from the previous question?

**A**

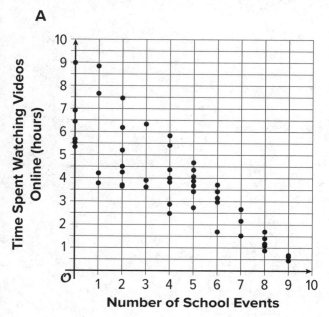

**B**

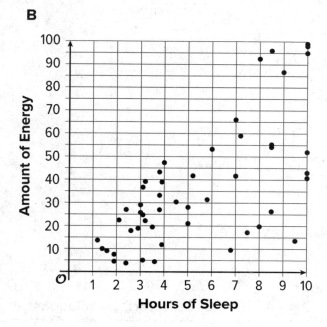

**C**

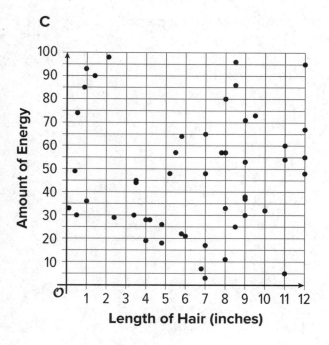

**D**

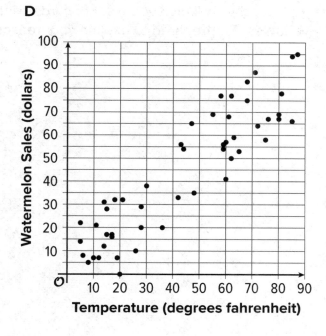

## Activity

### 8.3 Card Sort: Correlations

Your teacher will give you a set of cards. Match each scatter plot with a pair of variables. Be prepared to explain your reasoning.

**Lesson 3-9**

# What's the Correlation?

NAME _____ DATE _____ PERIOD _____

**Learning Goal** Let's reason about correlation of two variables in a situation.

## Warm Up
### 9.1 Which One Doesn't Belong: Correlations

Which one doesn't belong?

A. the number of pictures painted and the amount of paint left in the paint can

B. amount of ice cream eaten the previous summer and number of movies seen this summer

C. distance run and number of water breaks during the run

D. number of people who contracted a genetic disease and presence of the gene that raises risk for the disease

# Activity

## 9.2 Correlation Relationships

For each pair of graphs, the linear model fits the data about the same. What do you notice about the variables? How might the variables be related?

**1a.** the number of cows in some states and the number of chickens in those same states

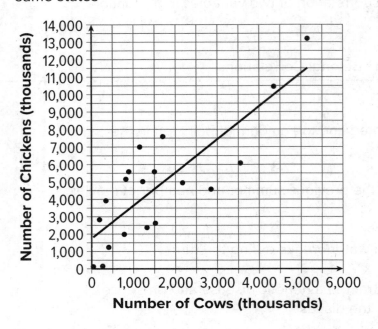

**1b.** the number of cows in some states and the number of farms in those same states

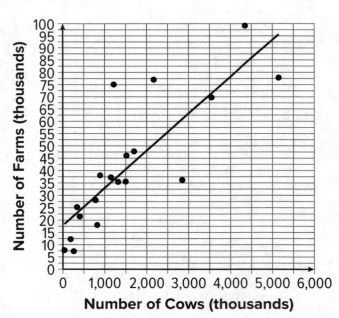

NAME _____ DATE _____ PERIOD _____

**2a.** the worth of a person's house and the worth of that same person's car

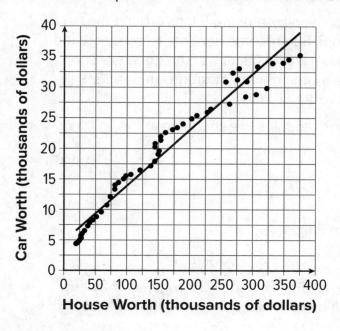

**2b.** the worth of a person's car and their income

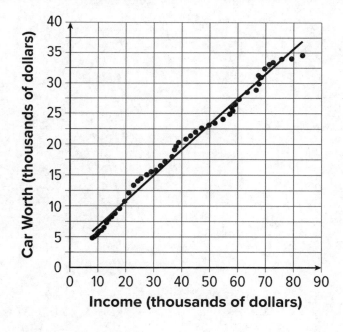

## Activity

### 9.3 It Takes Two

Mai is training for the upcoming track season by running 8 laps around the school track each morning before school. She records her time to complete the 8 laps and notices that she is finishing faster and faster as time goes on. She also notices that she feels better in the morning and her grades in her first class are improving as her times improve.

1. In addition to the 2 listed, what other variables are changing in this situation?

    a. time to complete 8 laps

    b. number of mornings Mai has run 8 laps

2. Select 3 pairs of variables from the list. For each pair determine if they are related, then decide whether you think one variable causes the other to change. Explain your reasoning.

    a.

    b.

    c.

Lesson 3-10

# Putting It All Together

NAME _____ DATE _____ PERIOD _____

**Learning Goal** Let's interpret data.

## Warm Up
**10.1 Which One Doesn't Belong: Data Correlations**

Which one doesn't belong?

A
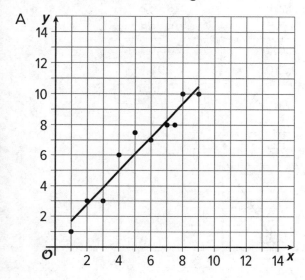

B
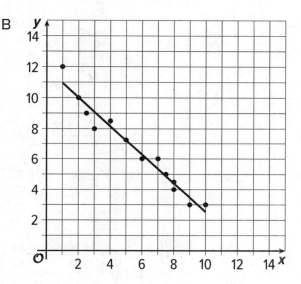

C
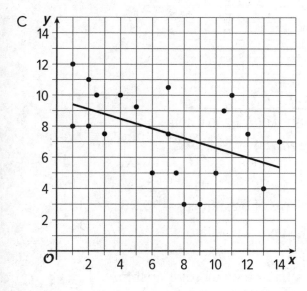

D

| x | y |
|------|-------|
| 3 | 6 |
| 3.75 | 8.50 |
| 7.25 | 7.50 |
| 5.50 | 11 |
| 6 | 9 |
| 8 | 10.25 |

Here are Elena's representations of the data set.

| Energy (kwh) | Electric Bill Price (dollars) |
|---|---|
| 500 | 50 |
| 560 | 57.60 |
| 610 | 65.10 |
| 675 | 70.25 |
| 700 | 74.80 |
| 755 | 90.66 |
| 790 | 92.34 |
| 836 | 105 |
| 892 | 150 |
| 940 | 173 |
| 932 | 182 |

| Energy (kwh) | Electric Bill Price (dollars) |
|---|---|
| 967 | 170 |
| 999 | 198 |
| 1,005 | 201.22 |
| 1,039 | 215.35 |
| 1,057 | 217 |
| 1,100 | 233 |
| 1,191 | 284.62 |
| 1,150 | 256.98 |
| 1,200 | 289.60 |
| 1,270 | 292 |

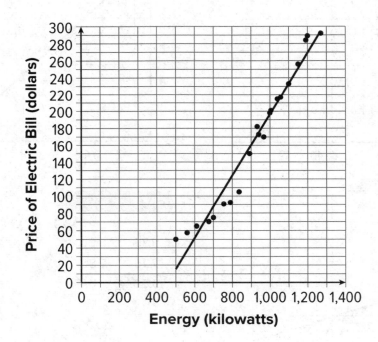

After analyzing the data, Elena concludes:

1. An estimate for the correlation coefficient for the line of best fit is $r = -0.98$.

2. Energy consumption and the price of electric bills have a positive relationship.

3. Energy consumption and the price of electric bills have a weak relationship.

4. Using the linear model, the electric bill is $260 when 1,200 kWh are consumed.

NAME _____ DATE _____ PERIOD _____

What parts of Elena's interpretation of the data do you agree with and what parts do you disagree with? Explain your reasoning.

## Activity

### 10.3 Confident Players

Before Diego's game, his coach asked each of his players, "On a scale of 1–10, how confident are you in the team winning the game?" Here is the data he collected from the team.

| players | Confidence in Winning (1–10) | Number of Points Scored in A Game |
|---------|------------------------------|-----------------------------------|
| Player A | 3 | 2 |
| Diego | 6 | 10 |
| Player B | 10 | 2 |
| Player C | 4 | 10 |
| Player D | 7 | 13 |
| Player E | 5 | 6 |
| Player F | 8 | 15 |
| Player G | 4 | 3 |
| Player H | 9 | 15 |
| Player I | 7 | 12 |
| Player J | 1 | 0 |
| Player K | 9 | 14 |
| Player L | 8 | 13 |
| Player M | 5 | 8 |

1.  Use technology to create a scatter plot, a line of best fit, and the correlation coefficient.

2.  Is there a relationship between players' level of confidence in winning and the amount of points they score in a game? Explain your reasoning.

3.  How many points does the linear model predict a player will score when his or her confidence is at a 4?

4.  Which players performed worse than the model predicted?

5.  Did Diego score better or worse than the linear model predicts?

# Functions

The total distance traveled by a drone can be found using an absolute value function. You will learn more about absolute value functions in this unit.

## Topics

- Functions and Their Representations
- Analyzing and Creating Graphs of Functions
- A Closer Look at Inputs and Outputs
- Inverse Functions
- Putting it All Together

Unit 4

# Functions

## Functions and Their Representations

## Analyzing and Creating Graphs of Functions

## A Closer Look at Inputs and Outputs

## Inverse Functions

## Putting it All Together

Lesson 4-1

# Describing Graphs

NAME _____ DATE _____ PERIOD _____

**Learning Goal** Let's describe graphs

## Warm Up

### 1.1 Notice and Wonder: A Rocket Path

A rocket is shot into the air and some aspects of its flight are shown in the graph.

What do you notice? What do you wonder?

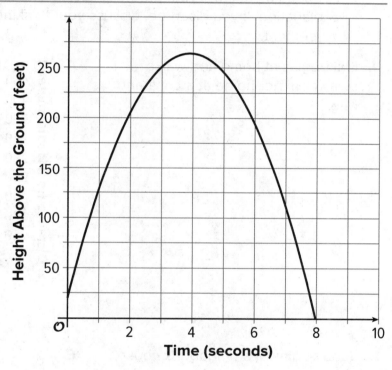

# Activity

### 1.2 Matching Descriptions and Graphs

Match the graph to the description of the situation.

Match each description in column A with a graph from column B that represents the situation. Be prepared to explain your reasoning.

1. Take turns with your partner to match a description with a graph.

   a. For each match that you find, explain to your partner how you know it's a match.

   b. For each match that your partner finds, listen carefully to their explanation. If you disagree, discuss your thinking and work to reach an agreement.

1. Mai begins at home and walks away from her home at a constant rate.

**A**

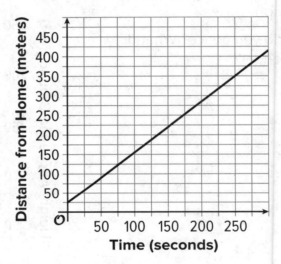

2. Jada begins at a neighbor's house and walks away from home at a constant rate.

**B**

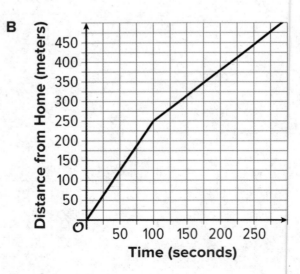

NAME _____ DATE _____ PERIOD _____

3. Clare begins her walk at school and walks home at a constant rate.

**C**

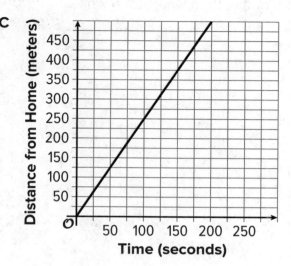

4. Elena begins at home and runs away from her home at a constant rate.

**D**

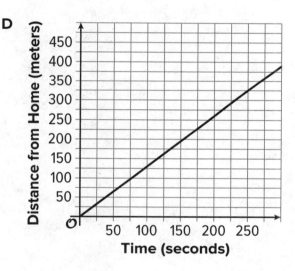

5. Lin begins at home and walks away from home for a while, then walks back home.

**E**

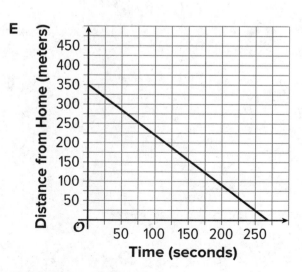

**6.** Priya begins at home and runs away
from home, then walks for a while.

**F**

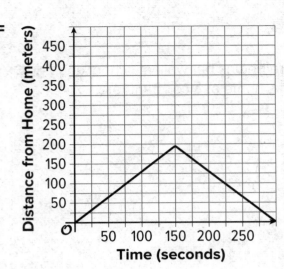

NAME _____ DATE _____ PERIOD _____

## Activity

### 1.3 Say What You See

In your own words, describe these graphs.

**1.**

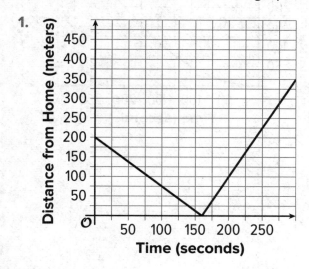

**2.**

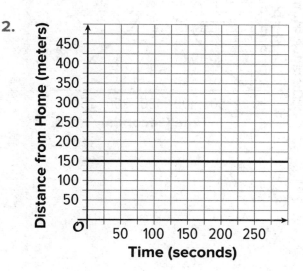

**3.**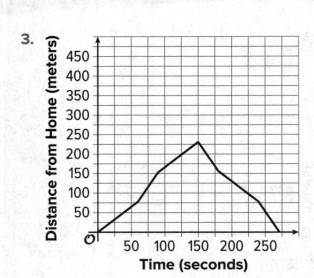

**4.**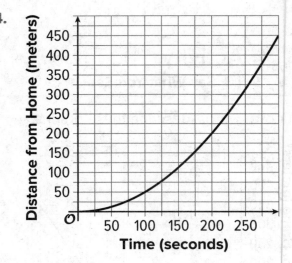

Lesson 4-2

# Understanding Points in Situations

NAME _____ DATE _____ PERIOD _____

**Learning Goal** Let's understand points on a function in a situation.

## Warm Up
### 2.1 A Day of Temperature

The temperature for a city is a function of time after midnight. The graph shows the values on a particular spring day.

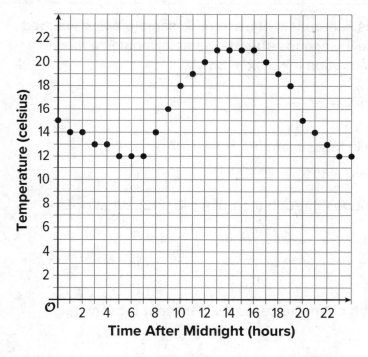

1. What does the point on the graph where $x = 15$ mean?

2. What is the temperature at 5 p.m.?

3. What is the hottest it gets on this day?

4. What is the coldest it gets on this day?

## Activity

### 2.2 What Happens to −2?

For each of these equations, find the value of $y$ when $x = -2$.

1. $y = 3x - 4$

2. $y = 10 - 2x$

3. $y = \frac{3}{2}x + 5$

4. $y = 2(x-1) + 4$

5. $y = -x + 19$

6. $y = \frac{x-3}{8}$

7. $y = 0.3x + 5$

## Activity

### 2.3 It's Heating Up!

The temperature, in degrees Fahrenheit, of a scientific sample being warmed steadily as a function of time in seconds after the sample is put in a machine can be represented by the equation $y = 2.1x + 86$.

1. What does it mean when $x = 2$?

2. What is the temperature in that situation?

3. What does it mean when $y = 122$?

4. A graph of this equation goes through the point (60, 212). What does that mean?

5. Give 2 values for $x$ that do not make sense. Explain your reasoning.

6. Give 2 values for $y$ that do not make sense. Explain your reasoning.

**Lesson 4-3**

# Using Function Notation

NAME _____ DATE _____ PERIOD _____

**Learning Goal** Let's use function notation to talk about points.

## Warm Up

**3.1 Which One Doesn't Belong: Function Notation**

Which one doesn't belong?

**A.** $f(0) = 2$

**B.** $(0, 5)$

**C.** $y = x + 2$

**D.**

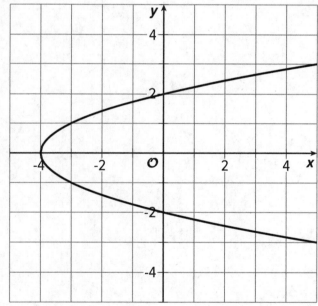

## Activity

### 3.2 Points into Function Notation and Back

1. A function is given by the equation $y = f(x)$. Write each of these coordinate pairs in function notation.

   a. $(2, 3)$

   b. $(-1, 4)$

   c. $(0, 3)$

   d. $(4, 0)$

   e. $\left(\dfrac{2}{3}, \dfrac{3}{4}\right)$

2. A function is given by the equation $h(x) = 5x - 3$. Write the coordinate pair for the point associated with the given values in function notation.

   a. $h(3)$

   b. $h(-4)$

   c. $h\left(\dfrac{2}{5}\right)$

NAME _____ DATE _____ PERIOD _____

## Activity

### 3.3 A Graph with Properties

**1.** Draw a graph of function $y = g(x)$ that has these properties:

- $g(0) = 2$
- $g(1) = 3$
- $(2, 3)$ is on the graph
- $g(5) = -1$

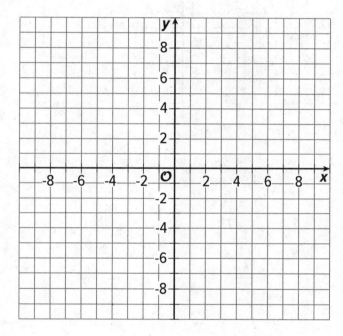

**2.** Han draws this graph for $g(x)$. What is the error?

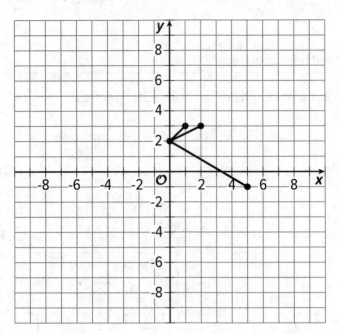

Lesson 4-4

# Interpreting Functions

NAME _____ DATE _____ PERIOD _____

**Learning Goal** Let's interpret some functions.

 Warm Up

**4.1 Math Talk: Finding Outputs**

Mentally evaluate the output for the input of 3.

$f(x) = 4\left(x - \frac{1}{2}\right)$

$g(x) = 2(6 - x)$

$h(x) = \frac{5}{3}x + \frac{1}{3}$

$j(x) = 0.2x - 1$

## Activity

### 4.2 It's Getting Hotter

A machine in a laboratory is set to steadily increase the temperature inside. The temperature in degrees Celsius inside the machine after being turned on is a function of time, in seconds, given by the equation $f(t) = 22 + 1.3t$.

1. What does $f(3)$ mean in this situation?

2. Find the value of $f(3)$ and interpret that value.

3. What does the equation $f(t) = 35$ mean in this situation?

4. Solve the equation to find the value of $t$ for the previous question.

5. Write an equation involving $f$ that represents each of these situations:

   a. The temperature in the machine 30 seconds after it is turned on.

   b. The time when the temperature inside the machine is 100 degrees Celsius.

NAME _____ DATE _____ PERIOD _____

## Activity
### 4.3 You Charge How Much?

Two companies charge to rent time using their supercomputers. Their fees are given by the equations $f(t) = 500 + 100t$ and $g(t) = 300 + 150t$. The lines $y = f(t)$ and $y = g(t)$ are graphed.

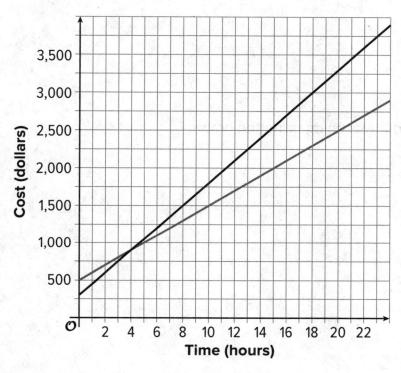

1. Which line represents $y = f(t)$? Explain how you know.

2. The lines intersect at the point (4, 900). What does this point mean in this situation?

3. Which is greater, $f(10)$ or $g(10)$? What does that mean in this situation?

**4.** Your group has $1,500 to spend on supercomputer time. Which company should your group use?

    **a.** Explain or show your reasoning using the equations.

    **b.** Explain or show your reasoning using the graph.

Lesson 4-5

# Function Representations

NAME _____ DATE _____ PERIOD _____

**Learning Goal** Let's examine different representations of functions.

 ## Warm Up

### 5.1 Notice and Wonder: Representing Functions

What do you notice? What do you wonder?

$$f(x) = \frac{2}{3}x - 1$$

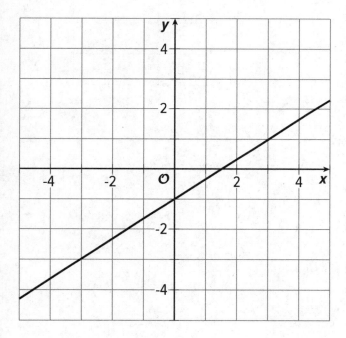

| x | y |
|----|------|
| -1 | $-\frac{5}{3}$ |
| 0 | -1 |
| 1 | $-\frac{1}{3}$ |
| 2 | $\frac{1}{3}$ |
| 3 | 1 |

# Activity

## 5.2 A Seat at the Tables

Use the equations to complete the tables.

**1.** $y = 3x - 2$

| x | y |
|---|---|
| 1 | |
| 3 | |
| -2 | |

**3.** $y = \frac{1}{2}x + 2$

| x | y |
|---|---|
| -4 | |
| 3 | |
| 6 | |

**2.** $y = 5 - 2x$

| x | y |
|---|---|
| 0 | |
| 3 | |
| 5 | |

**4.**

| x | $y = 2x - 10$ |
|---|---|
| 3 | |
| 7 | |
| -8 | |

NAME _____ DATE _____ PERIOD _____

# Activity
## 5.3 Function Finder

1. Use the values in the table to graph a possible function that would have the values in the table.

a.

| x | y |
|---|---|
| 1 | 3 |
| 2 | 5 |
| 3 | 7 |
| 5 | 11 |

b.

| x | y |
|---|---|
| -2 | 0 |
| 0 | 1 |
| 2 | 2 |
| 4 | 3 |

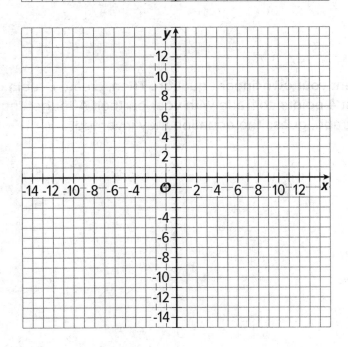

**c.**

| x | y |
|---|---|
| -2 | 14 |
| -1 | 12 |
| 1 | 8 |
| 2 | 6 |

**2.** For each of the tables and graphs, write a linear equation (like $y = ax + b$) so that the table can be created from the equation.

**3.** Invent your own linear equation. Then, create a table or graph, including at least 4 points, to trade with your partner. After getting your partner's table or graph, guess the equation they invented.

Lesson 4-6

# Finding Interesting Points on a Graph

NAME _____ DATE _____ PERIOD _____

**Learning Goal** Let's find some interesting points.

## Warm Up
### 6.1 Notice and Wonder: Unemployment Percentage

What do you notice? What do you wonder?

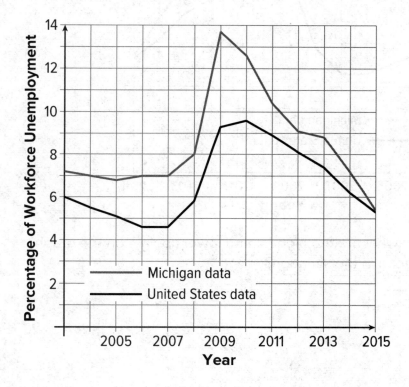

This graph shows the percentage of the workforce that is unemployed in the United States and Michigan for several years.

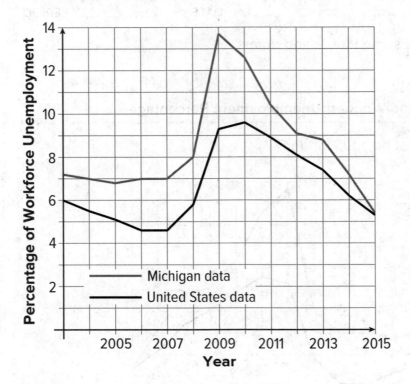

1. For the United States, what are the highest and lowest points on the graph? What do the points mean in the situation?

2. For Michigan, what are the highest and lowest points on the graph? What do the points mean in the situation?

NAME _____ DATE _____ PERIOD _____

## Activity
### 6.3 The Wire

1. Use technology to graph the function
   $f(x) = x^4 - 16x^3 + 86x^2 - 176x + 105$.

2. What are some points on the graph that you think are interesting? Explain your reasoning.

**3.** Examine the graph representing electrical voltage in a wire as a function of time. What interesting points do you see? Explain your reasoning.

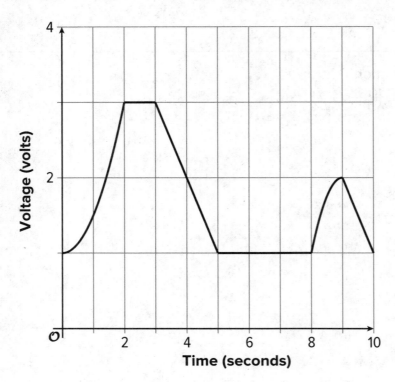

**4.** Use the points you found to describe what is happening to the voltage within the wire.

Lesson 4-7

# Slopes of Segments

NAME _____ DATE _____ PERIOD _____

**Learning Goal** Let's look at slopes again.

## Warm Up
### 7.1 Math Talk: Evaluating Fractions

Evaluate mentally.

$\dfrac{102 - 96}{45 - 42}$

$\dfrac{-8 - 4}{6 - 2}$

$\dfrac{31 - 18}{5 - 10}$

$\dfrac{4 - 9}{12 - 18}$

1.  Find the slope of the line that connects the given points.

    a.  (0, 0) and (3, 2)

    b.  (4, 2) and (10, 7)

    c.  (1, -2) and (2, 5)

    d.  (-3, 4) and (-5, -2)

    e.  (8, 3) and (10, -9)

2.  For each pair of points, find the slope of the line that goes through the 2 points.

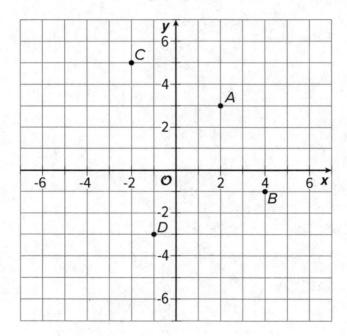

    a.  *A* and *B*

    b.  *A* and *D*

    c.  *B* and *C*

    d.  *C* and *D*

NAME _____ DATE _____ PERIOD _____

## Activity

### 7.3 Ups and Downs

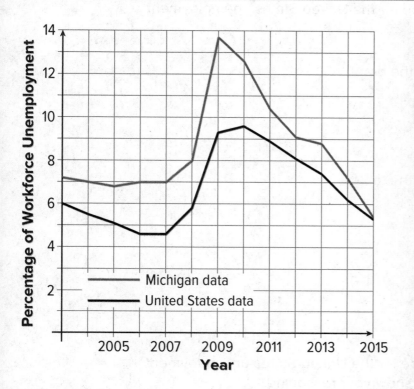

| Year | Michigan | United States |
|------|----------|---------------|
| 2003 | 7.2 | 6 |
| 2004 | 7 | 5.5 |
| 2005 | 6.8 | 5.1 |
| 2006 | 7 | 4.6 |
| 2007 | 7 | 4.6 |
| 2008 | 8 | 5.8 |
| 2009 | 13.7 | 9.3 |
| 2010 | 12.6 | 9.6 |
| 2011 | 10.4 | 8.9 |
| 2012 | 9.1 | 8.1 |
| 2013 | 8.8 | 7.4 |
| 2014 | 7.2 | 6.2 |
| 2015 | 5.4 | 5.3 |

1. What do the slopes of the segments mean?

**2.** Find the slope of the segment between 2004 and 2005 for unemployment in Michigan.

**3.** Between what 2 years is the slope for the United States unemployment percentage greatest?

    **a.** Explain your reasoning using the graph.

    **b.** Explain your reasoning using the table.

**4.** Between what 2 years is the slope for the United States unemployment percentage the least? Explain or show your reasoning.

Lesson 4-8

# Interpreting and Drawing Graphs for Situations

NAME _____ DATE _____ PERIOD _____

**Learning Goal** Let's make sense of graphs and scenarios.

 Warm Up
### 8.1 Notice and Wonder: Crimes

What do you notice? What do you wonder?

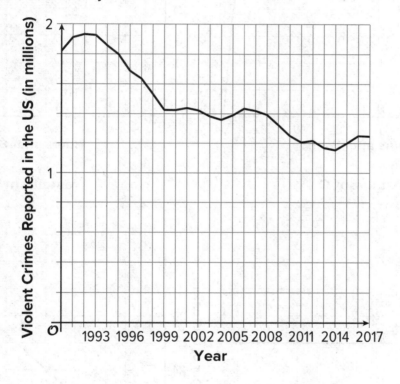

These graphs show how busy restaurants are at different times of the day.

### restaurant A

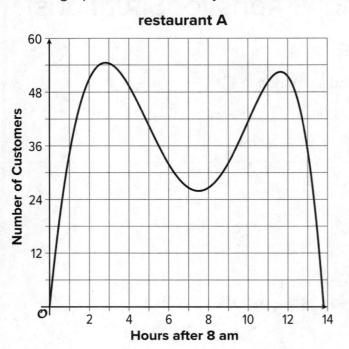

### restaurant B

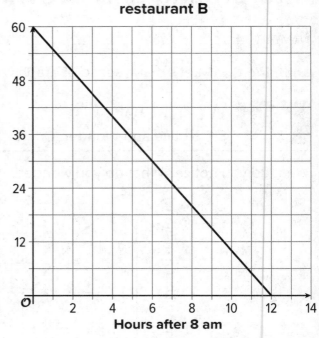

### restaurant C

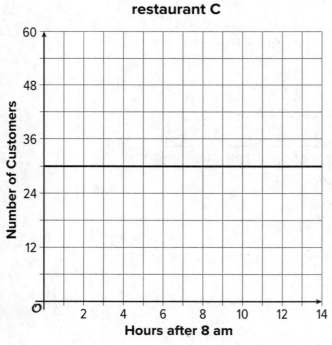

### restaurant D

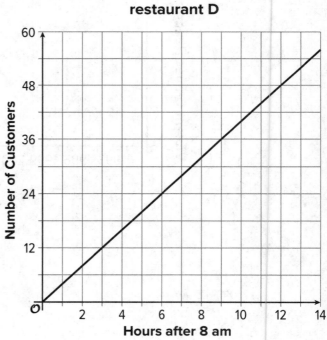

NAME _____ DATE _____ PERIOD _____

For each situation, select the best restaurant. Be prepared to explain your reasoning.

1. Which restaurant is busy in the morning, then has fewer customers in the evening?

2. If Lin's mom wants to go to a popular dinner restaurant, which restaurant should Lin take her mom to eat?

3. Noah's dad prefers breakfast places with few customers so that he can start on work while eating. Which restaurant should Noah's dad go to for breakfast?

4. Which restaurant would you visit during a 30 minute lunch break? 1 hour lunch break?

## Activity

### 8.3 Draw the Graphs

For each situation, draw a graph that could represent it.

1. Diego starts at home and walks away from home at a steady rate of 3 miles per hour.

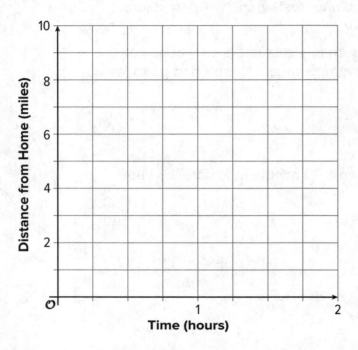

2. Mai starts 5 miles from home and walks at a steady rate of 3 miles per hour toward her home until she gets there and stays.

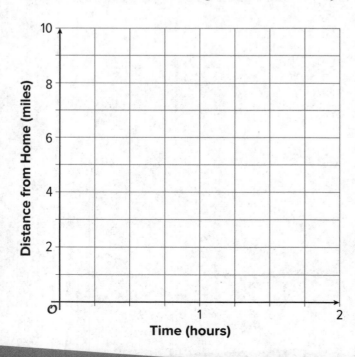

NAME _____ DATE _____ PERIOD _____

**3.** A soccer player kicks a ball that's on the ground so that it goes up to a height of about 10 feet and then comes back down to hit the ground 1.55 seconds later.

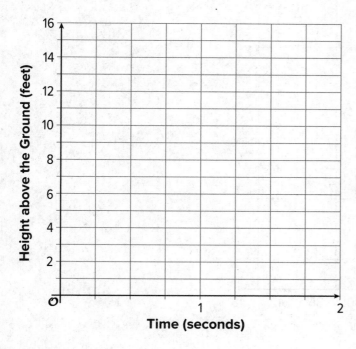

**4.** The amount of charge left in a phone battery as a percentage is a function of time. Clare runs her phone until it is completely dead, then charges it all the way back up at a steady rate.

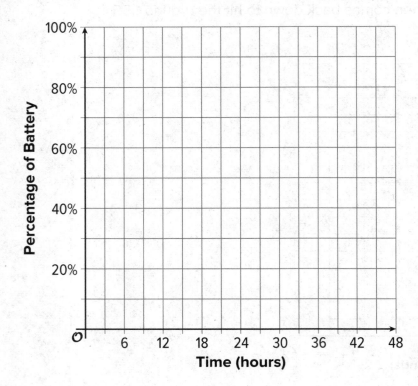

Lesson 4-9

# Increasing and Decreasing Functions

NAME _____ DATE _____ PERIOD _____

**Learning Goal** Let's look at what a graph does based on a situation.

 ## Warm Up
### 9.1 Comparing Values

For each pair of numbers, write = , <, or > in the blank to make a true equation or inequality. Be prepared to share your reasoning.

1. -6 _____ -9

2. $\dfrac{7}{3}$ _____ $\dfrac{13}{6}$

3. 5.2 _____ $\dfrac{53}{11}$

4. $5(3 - 6)$ _____ $15 - 6$

5. Let $f(x) = 5 - 2x$.

   a. $f(3)$ _____ $f(5)$

   b. $f(-3)$ _____ $f(-4)$

   c. $f(-1)$ _____ $f(1)$

### 9.2 What Could It Be?

Describe $f(x)$ and $g(x)$ with a situation that could fit the given graphs. Explain
your reasoning.

1.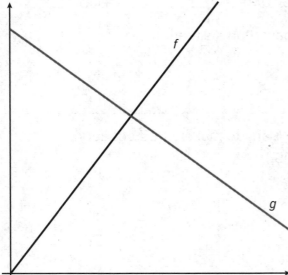

2.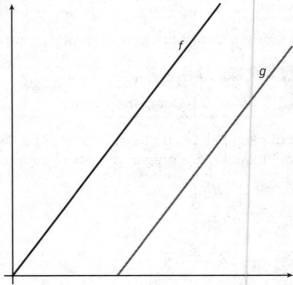

NAME _____ DATE _____ PERIOD _____

**3.**

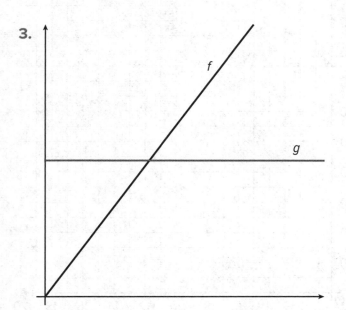

**4.**

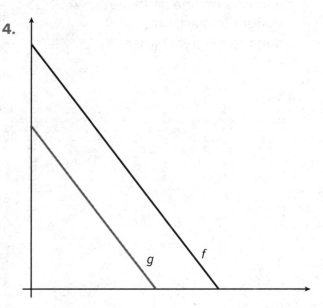

 ## Activity

### 9.3 Cities, Towns, and Villages

Draw an example of a graph that shows two functions as they are described. Make sure to label the functions.

**1.** The population of 2 cities as functions of time so that city A always has more people than city B.

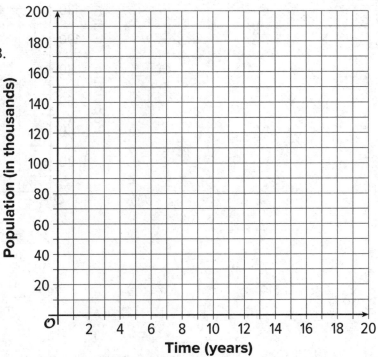

**2.** The population of 2 towns as functions of time so that town A is larger to start, but then town B gets larger.

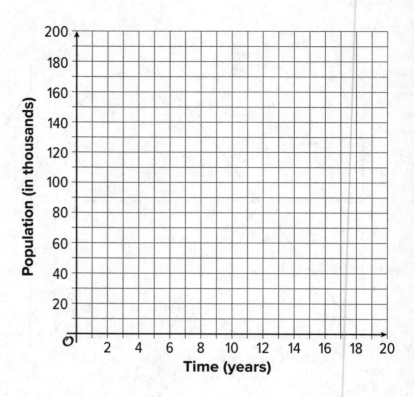

**3.** The population of 2 villages as functions of time so that village A has a steady population and village B has a population that is initially large, but decreases.

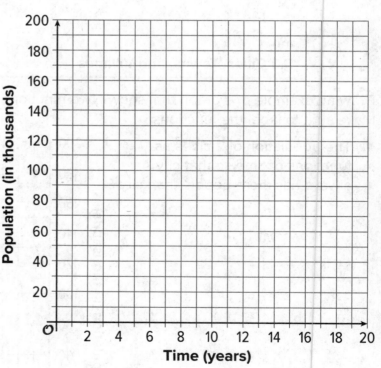

Lesson 4-10

# Interpreting Inputs and Outputs

NAME _____ DATE _____ PERIOD _____

**Learning Goal** Let's look at inputs and outputs of a function.

## Warm Up
### 10.1 A Function Riddle

The table shows inputs and outputs for a function. What function could it be?

| Input | Output |
|-------|--------|
| 1 | 3 |
| 2 | 3 |
| 3 | 5 |
| 4 | 4 |
| 5 | 4 |
| 10 | 3 |
| 11 | 6 |

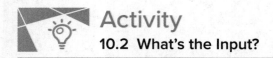
1. For each pair of variables, which one makes the most sense as the input? When possible, include a reasonable unit.

   a. The number of popcorn kernels left unpopped as a function of time cooked.

   b. The cost of crab legs as a function of the weight of the crab legs.

   c.

   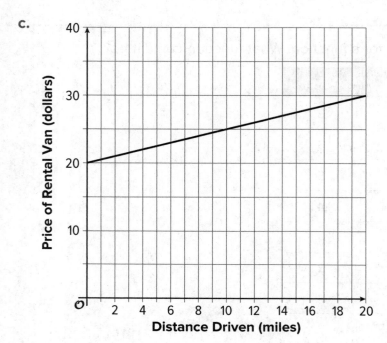

   d.

   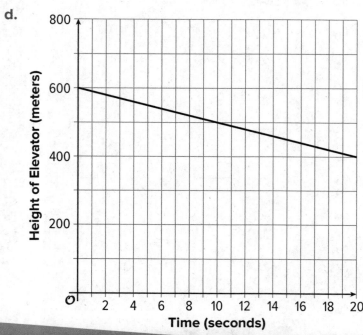

e. $f(t) = 5t + 8$ where $t$ represents the time that a bike is rented, in hours, and $f(t)$ gives the cost of renting the bike.

f. $g(n) = 7n + 4$ where $n$ represents the number of pencils in a box and $g(n)$ represents the weight of the box of pencils in grams.

2. Write the equation or draw the graph of a function relating the 2 variables.

   a. Input: side length of a square, output: perimeter of the square

   b. Input: time spent walking (minutes), output: distance walked (meters)

   c. Input: time spent working out (minutes), output: heart rate (beats per minute)

# Activity

## 10.3 Matching Possible Inputs

For each function in column A, find which inputs in column B could be used in the function. Be prepared to explain your reasoning for whether you include each input or not.

1. Take turns with your partner to match a function with its possible inputs.

    a. For each function, explain to your partner whether each input is possible to use in the function or not.

    b. For each input, listen carefully to their explanation. If you disagree, discuss your thinking and work to reach an agreement.

1. $f$(person) = the person's birthday

2. $g(x) = 2x + 1$

3. $h$(item) = the number of chromosomes in the item

4. $P$(equilateral triangle side length) = 3 · (side length)

5. $C$(number of students) = 9.99 (number of students) + 15

- Martha Washington (the first First Lady of the United States)
- an apple
- 6
- 9.2
- 0
- -1

For each function, write 2 additional inputs that make sense to use. Write 1 additional input that does not make sense to use. Be prepared to share your reasoning.

Lesson 4-11

# Examining Domains and Ranges

NAME _____ DATE _____ PERIOD _____

**Learning Goal** Let's play with graphs, domains, and ranges in situations.

## Warm Up
### 11.1 Notice and Wonder: A Wiggly Graph

What do you notice? What do you wonder?

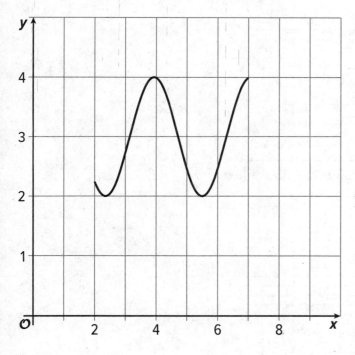

 Activity

**11.2 Moving Weeds and a Ball**

**1.** Examine these graphs and describe a situation that could match the situation.

**a.**

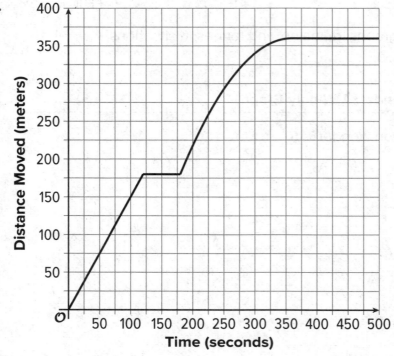

**b.**

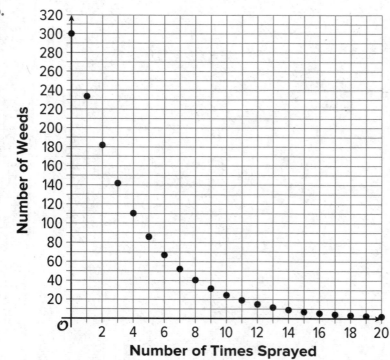

NAME _____ DATE _____ PERIOD _____

**c.**

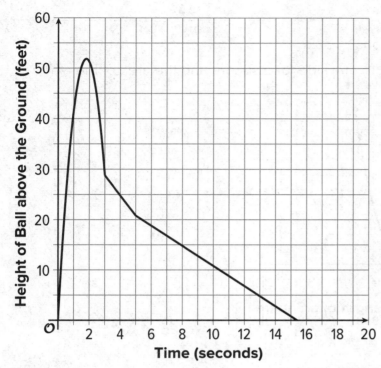

**2.** For each situation, give an example of a value that could be:

• In the domain

• Not in the domain

• In the range

• Not in the range

1. What is wrong with these graphs?

   a. The graph relates the length of a side for a square and the area of the square.

   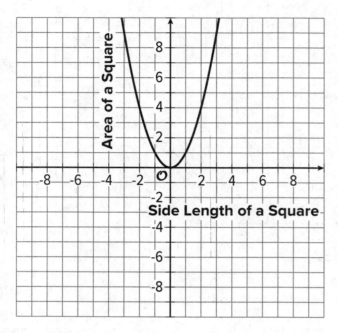

   b. The graph relates the number of students going on a field trip and the cost of the trip.

   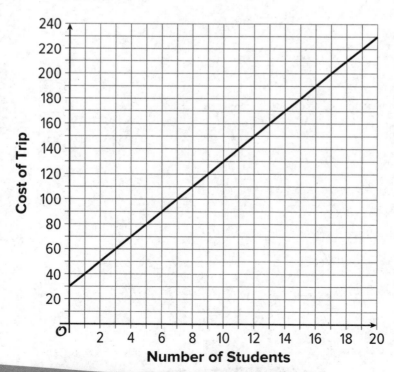

NAME _____ DATE _____ PERIOD _____

c. The graph represents Han's height since he was 4 years old until now when he is 14.

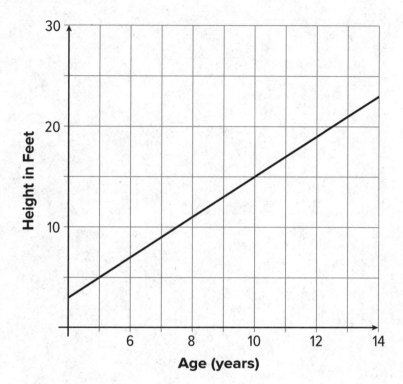

2. On each graph, draw a more realistic graph.

Lesson 4-12

# Functions with Multiple Parts

NAME _____ DATE _____ PERIOD _____

Learning Goal Let's look at domains that have boundaries.

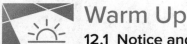

## Warm Up
### 12.1 Notice and Wonder: Ticket Price

What do you notice? What do you wonder?

| Age | Ticket Cost |
|---|---|
| 0–2 | FREE |
| 2–5 | $2.00 |
| 5–12 | $5.00 |
| 13–16 | $7.50 |
| 17–50 | $10.00 |
| 55 and up | $5.00 |

# Activity
## 12.2 Group Ticket Cost

A community orchestra charges different amounts for tickets to shows based on the age of the person attending. A sign in front of the box office where tickets are sold shows the prices.

| Age | Ticket Cost |
|---|---|
| 0–2 | FREE |
| 3–13 | $4.00 |
| 14–18 | $6.00 |
| 19–25 | $9.00 |
| 26–54 | $10.00 |
| 55 and up | $6.00 |

1. How much does each group need to pay for their tickets?

   a. 2 adults aged 40 and 36, and 2 kids aged 4 and 1

   b. 3 adults aged 74, 37, and 36

   c. 5 adults in their 30s and 25 students aged 15 and 16

   d. 1 adult aged 25 and 4 kids aged 1, 9, 13, and 16

2. A mother arrives and tells the box office clerk that her child is 35 months old. How much should the clerk charge for the child?

3. If there is a rule that uses the age of a person attending the orchestra concert as the input and outputs the ticket price for that person, is that rule a function? Explain your reasoning.

   a. What is the domain for the rule?

   b. What is the range for the rule?

NAME _____ DATE _____ PERIOD _____

## Activity

### 12.3 A Light Trip

1. Noah leaves his home, sometimes running, sometimes walking, sometimes stopping until he remembers that he doesn't have his wallet, then he goes back home. A graph representing his journey is shown in the graph.

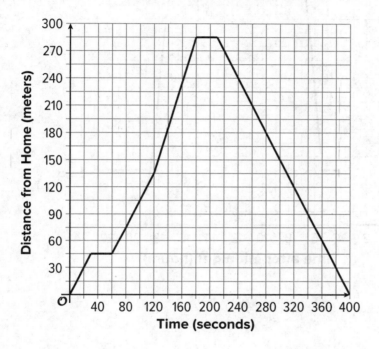

**a.** Describe what is happening on the domain $210 < x < 400$.

**b.** What are the domain intervals that represent the times when Noah was running?

**c.** What are the domain intervals that represent the times when Noah was stopped?

**d.** What are the domain intervals that represent the times when Noah was walking away from home?

**2.** The amount of light in a room is shown as a function of the number of hours after midnight. Describe what might be happening in the room. Be sure to use intervals within the domain in your description.

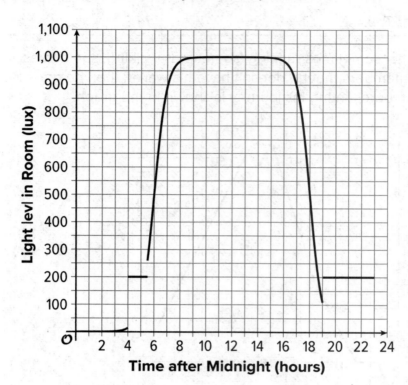

Lesson 4-13

# Number Line Distances

NAME _____ DATE _____ PERIOD _____

**Learning Goal** Let's calculate distances between numbers.

## Warm Up
### 13.1 Math Talk: How Far?

Evaluate mentally: How far away is each house from the school?

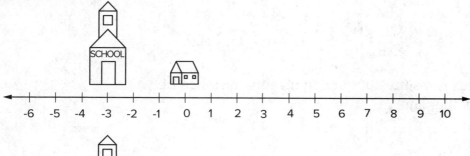

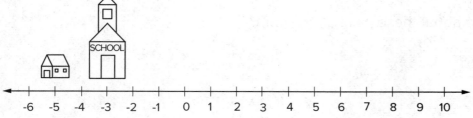

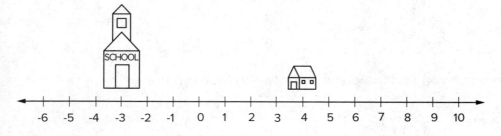

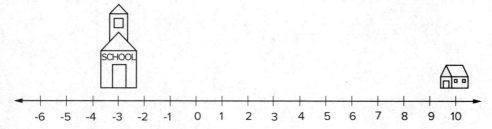

1. For each pair of values, find $b - a$. Be prepared to explain your reasoning.

   a. $a = 28, b = 57$

   b. $a = \frac{4}{5}, b = \frac{1}{2}$

   c. $a = 27, b = -17$

   d. $a = -35, b = -19$

   e. $a = 19, b = 35$

   f. $a = -106, b = 43$

2. For which pairs of values does the subtraction give the distance between the numbers on the number line?

   a. What do you notice about these pairs of numbers?

3. Given 2 numbers, how can you find the distance between them on the number line?

NAME _____ DATE _____ PERIOD _____

 Activity

**13.3 It's That Far Away**

1. Find 2 numbers that are $d$ away from $a$ on the number line.

   a. $a = 14, d = 6$

   b. $a = -7, d = 16$

   c. $a = 103, d = 56$

   d. $a = 4, d = 138$

2. Use $d$ and $a$ to write 2 expressions that find the values that are $d$ away from $a$.

3. Kiran is looking at some old work where he did problems like this and found an answer that was marked correct. The answer is -18 and 46. Could Kiran figure out the values of $a$ and $d$ from the problem based on these values? If so, what are the values? If not, what additional information would help? Explain or show your reasoning.

4. In a planned neighborhood along Stepford Street, all of the houses are identical and equally distant from one another. The house at 102 Stepford Street is 2,250 feet from the house at 84 Stepford Street. Is there enough information to find the address of another house that is that same distance away from 84 Stepford Street? Explain your reasoning.

Lesson 4-14

# Absolute Value Meaning

NAME _____ DATE _____ PERIOD _____

**Learning Goal** Let's investigate absolute values.

 **Warm Up**

**14.1 Math Talk: Closest to Zero**

For each pair of values, decide mentally which one has a value that is closer to 0.

$-\frac{4}{5}$ or $\frac{5}{4}$

$\frac{1}{12}$ or $\frac{1}{14}$

-0.001 or 0.0001

1.3 or $\frac{4}{3}$

## Activity

### 14.2 What is Absolute Value?

1. One of the lowest places in Europe is 23 feet below sea level. We can use "-23 feet" to describe its elevation, and "|-23| feet" to describe its vertical distance from sea level. In the context of elevation, what would each of the following numbers describe?

   a. 37 feet

   b. |37| feet

   c. -6 feet

   d. |-6| feet

2. Water freezes at 0 degrees Celsius. For each pair of temperatures, which temperature is closest to the freezing point of water? Use absolute value to explain your reasoning.

   a. 14°C or -15°C

   b. -5°C or -2°C

   c. -12°C or 18°C

## Activity

### 14.3 Absolute Value Expressions

Find the value of these expressions.

1. $|10 - 12|$                    4. $|8 - 10| - 16$

2. $|-16 + 5|$                    5. $2|-13 - 2| + 18$

3. $3|9 - 4|$                     6. $9 - |19 + 3|$

**Lesson 4-15**

# Finding Input Values and Function Values

NAME _____ DATE _____ PERIOD _____

**Learning Goal** Let's play with inputs and outputs of functions.

## Warm Up
### 15.1 Inches to Feet and Back

There are 12 inches in 1 foot.

Complete the table by converting the lengths to the other unit.

| Inches | Feet |
|--------|------|
| 36     |      |
| 18     |      |
|        | 4    |
|        | 6.3  |
| 105    |      |

# Activity

## 15.2 Building Square Planters

For a class project, students are building an outdoor square planter from pieces of wood, then filling it with soil. The amount of soil needed is based on the area within the square.

NAME _____ DATE _____ PERIOD _____

1. Mai has boards of wood that are 2 meters long. What is the area of the largest square planter she could make?

2. Tyler has boards of wood that are 5 feet long. What is the area of the largest square planter he could make?

3. Lin has boards of wood that are 53 inches long. What is the area of the largest square planter she could make?

4. Elena has enough soil to fill 36 square feet in a planter. What length boards should she cut to make the square planter that can hold all the soil?

5. Andre has enough soil to fill 16 square meters in a planter. What length boards should he cut to make the square planter that can hold all the soil?

6. If a student has boards of wood that are $s$ feet long, what is the area of the largest square planter they can build? Write the solution as an equation involving $f(s)$.

**15.3 Inputs and Outputs from Graphs**

The graph represents $y = f(x)$.

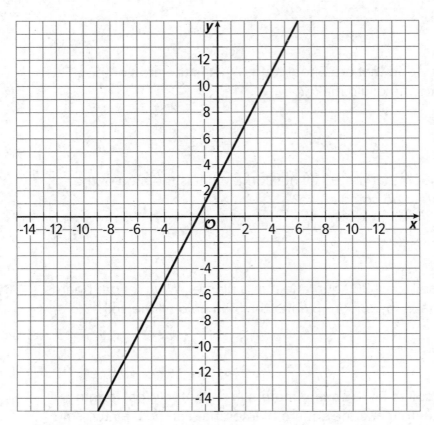

1. Use the graph to find the values.

   a. $f(2)$

   b. $f(-1)$

   c. $f(5)$

   d. $f(0)$

2. Use the graph to find the value of $x$ that makes these true.

   a. $f(x) = 11$

   b. $f(x) = 9$

   c. $f(x) = -1$

   d. $f(x) = 5$

Lesson 4-16

# Rewriting Equations for Perspectives

NAME _____ DATE _____ PERIOD _____

**Learning Goal** Let's match and rewrite linear equations.

## Warm Up
### 16.1 No Bad Apples

Which option would you select? Use mathematical reasoning to explain your selection.

Option A: Each apple costs $0.97 and are on sale with a "Buy 2, Get 1 Free" offer.

Option B: Bags of 6 apples are on sale "2 for $7.50" but you must buy 2 bags.

# Activity

## 16.2 A Charity Shopping Trip

A person has collected a lot of money for providing clothing to those in need. They go to a store to buy several clothing items with the money collected.

Match each description in column A with an equation from column B that represents the situation. Be prepared to explain your reasoning.

1. Take turns with your partner to match a description of a situation with an equation that represents the situation.

   a. For each match that you find, explain to your partner how you know it's a match.

   b. For each match that your partner finds, listen carefully to their explanation. If you disagree, discuss your thinking and work to reach an agreement.

NAME _____ DATE _____ PERIOD _____

1. A store charges $6 for each shirt sold. A person buys x shirts and pays y dollars for the total.

$$y = 6x$$

2. A store charges $6 for each pair of shorts sold. They also offer a $3 coupon to be used on the entire order. A person buys x pairs of shorts and pays y dollars for the total after using the coupon.

$$y = \frac{6x}{3}$$

3. A store charges $6 for 3 pairs of socks. A person buys x pairs of socks and pays y dollars for the total.

$$y = \frac{3x}{6}$$

4. A store charges $6 for each pair of shoes sold and also charges $3 to lace up all of the shoes in the entire order. A person buys x pairs of shoes and pays y for the total including lacing up all the shoes.

$$y = 3x - 6$$

5. A store charges $3 for 6 handkerchiefs. A person buys x handkerchiefs and pays y for the total.

$$y = 6x - 3$$

6. A store charges $3 for each pair of gloves sold. They also offer a $6 coupon to be used on the entire order when there are more than 4 pairs of gloves purchased. A person buys x pairs of gloves (with $x > 4$) and pays y dollars for the total after using the coupon.

$$y = 6x + 3$$

## Activity

**16.3 Isolate the x**

Rearrange the equations so that one side of the equation is only $x$. Be prepared to explain or show your reasoning.

1. $T = x - 2$

2. $T = 2x$

3. $T = 2x - 1$

4. $T = \dfrac{x}{2}$

5. $T = 2(x - 1)$

6. $T = \dfrac{x - 1}{2}$

**Lesson 4-17**

# Interpreting Function Parts in Situations

NAME _____ DATE _____ PERIOD _____

**Learning Goal** Let's pick apart functions

## Warm Up

### 17.1 Math Talk: Function Evaluation

Mentally find the value of $x$ for the given function value using the function: $f(x) = 3(x - 2)$

$f(x) = 9$

$f(x) = 210$

$f(x) = 10$

$f(x) = 0$

## Activity

### 17.2  A Long Car Trip

On a long car trip, the distance on the odometer (in miles) is a function of time (in hours after the trip begins) given by the equation $d(t) = 34t + 45{,}233$.

1. What is the rate of change for the function? What does it mean in this situation?

2. What is the value of $d(0)$? What does it mean in this situation?

3. What is the value of $d(-1)$? What does it mean in this situation?

4. When is $d(t) = 45{,}800$?

5. Do each of the values make sense? Explain your reasoning.

NAME _____ DATE _____ PERIOD _____

## Activity

### 17.3  A Warehouse and Highway

1. A warehouse in a factory initially holds 2,385 items and receives all of the items made in production throughout a day. During a particular day, the factory produces 150 items per hour to put into the warehouse. Write a function, *f*, to represent the number of items in the warehouse at time *t* after production begins for the day.

   **a.** What are the units for *t*?

   **b.** What is the domain of the function? Explain your reasoning.

   **c.** What is the range of the function? Explain your reasoning.

   **d.** What is the value of *t* when $f(t) = 3,000$? What does that mean in this situation?

2. During a focused effort on building new infrastructure for 3 years, a company can build 0.8 miles of highway per day. The company has already built 12 miles of highway before the focused effort. Write a function, $g$, to represent the length of highway built by the company as a function of $t$ during the focused effort.

   a. What are the units for $g(t)$?

   b. What is the domain of the function? Explain your reasoning.

   c. What is the range of the function? Explain your reasoning.

   d. What is the value of $t$ when $g(t) = 400$? What does that mean in this situation?

Lesson 4-18

# Modeling Price Information

NAME _____ DATE _____ PERIOD _____

**Learning Goal** Let's predict some information.

## Warm Up
### 18.1 What'll It Be?

The points on the graph represent the average resale price of a toy in dollars as a function of time.

1. Use the information to predict the average resale price of the toy on day 12. Explain your reasoning.

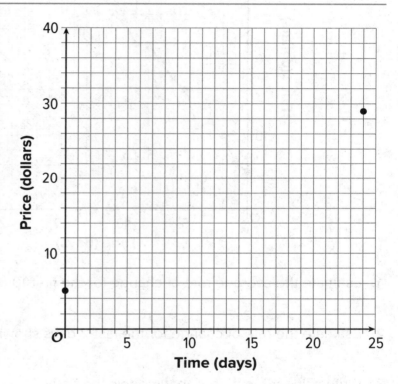

2. How confident are you in your predictions? Explain your reasoning.

### 18.2 Collectable Toy Price

The graph shows the average resale price for a toy in dollars as a function of time in days.

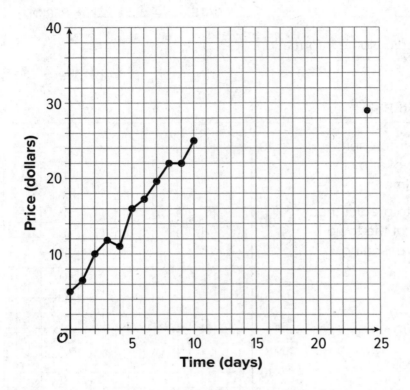

1. Estimate the average rate of change for the first 10 days.

2. Estimate the rate of change for the last 2 days shown.

3. Write a linear function, $f$, that models the data.

4. Predict the price of the toy after 12 days.

NAME _____ DATE _____ PERIOD _____

## Activity
### 18.3 More Information

After a few more days, a graph of the average price of the toy looks like this.

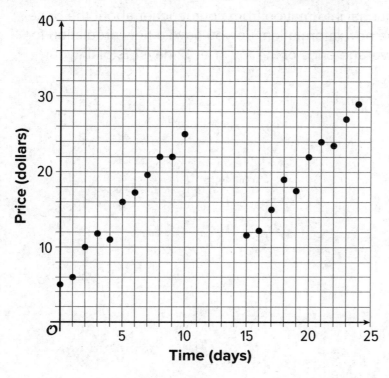

**1.** Draw a function (it does not need to be linear) that could model the data.

2. Use your graph to predict the average price of the toy after 12 days. How confident are you in this answer?

3. Pause here to get additional information from your teacher about the price of the toy. Based on the new information, do you have a new prediction for what happens to the average price of the toy after 12 days? Explain your reasoning.

# Glossary

**absolute value**  The absolute value of a number is its distance from 0 on the number line.

**absolute value function**  The function $f$ given by $f(x) = |x|$.

**association**  In statistics we say that there is an association between two variables if the two variables are statistically related to each other; if the value of one of the variables can be used to estimate the value of the other.

**average rate of change**  The average rate of change of a function $f$ between inputs $a$ and $b$ is the change in the outputs divided by the change in the inputs: $\frac{f(b) - f(a)}{b - a}$. It is the slope of the line joining $(a, f(a))$ and $(b, f(b))$ on the graph.

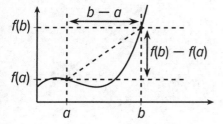

**bell-shaped distribution**  A distribution whose dot plot or histogram takes the form of a bell with most of the data clustered near the center and fewer points farther from the center.

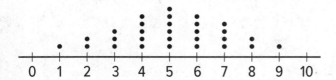

**bimodal distribution**  A distribution with two very common data values seen in a dot plot or histogram as distinct peaks. In the dot plot shown, the two common data values are 2 and 7,

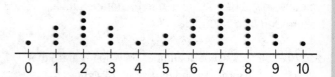

**categorical data**  Categorical data are data where the values are categories. For example, the breeds of 10 different dogs are categorical data. Another example is the colors of 100 different flowers.

**categorical variable**  A variable that takes on values which can be divided into groups or categories. For example, color is a categorical variable which can take on the values, red, blue, green, etc.

**causal relationship**  A relationship is one in which a change in one of the variables causes a change in the other variable.

**coefficient**  In an algebraic expression, the coefficient of a variable is the constant the variable is multiplied by. If the variable appears by itself then it is regarded as being multiplied by 1 and the coefficient is 1.

The coefficient of $x$ in the expression $3x + 2$ is 3. The coefficient of $p$ in the expression $5 + p$ is 1.

**completing the square**  Completing the square in a quadratic expression means transforming it into the form $a(x + p)^2 - q$, where $a$, $p$, and $q$ are constants.

Completing the square in a quadratic equation means transforming into the form $a(x + p)^2 = q$.

**constant term**  In an expression like $5x + 2$ the number 2 is called the constant term because it doesn't change when $x$ changes.

In the expression $5x - 8$ the constant term is -8, because we think of the expression as $5x + (-8)$. In the expression $12x - 4$ the constant term is -4.

constraint  A limitation on the possible values of variables in a model, often expressed by an equation or inequality or by specifying that the value must be an integer. For example, distance above the ground $d$, in meters, might be constrained to be non-negative, expressed by $d \geq 0$.

correlation coefficient  A number between -1 and 1 that describes the strength and direction of a linear association between two numerical variables. The sign of the correlation coefficient is the same as the sign of the slope of the best fit line. The closer the correlation coefficient is to 0, the weaker the linear relationship. When the correlation coefficient is closer to 1 or -1, the linear model fits the data better. The first figure shows a correlation coefficient which is close to 1, the second a correlation coefficient which is positive but closer to 0, and the third a correlation coefficient which is close to -1.

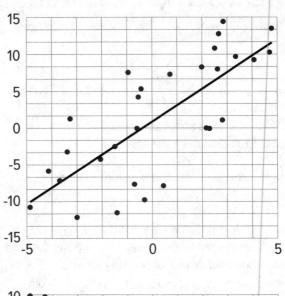

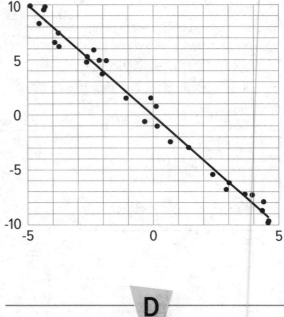

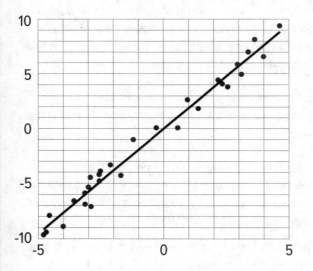

## D

decreasing (function)  A function is decreasing if its outputs get smaller as the inputs get larger, resulting in a downward sloping graph as you move from left to right.

A function can also be decreasing just for a restricted range of inputs. For example the function $f$ given by $f(x) = 3 - x^2$, whose graph is shown, is decreasing for $x \geq 0$ because the graph slopes downward to the right of the vertical axis.

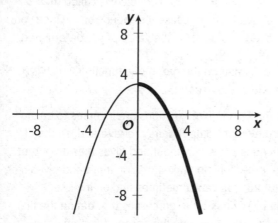

**dependent variable** A variable representing the output of a function.

The equation $y = 6 - x$ defines $y$ as a function of $x$. The variable $x$ is the independent variable, because you can choose any value for it. The variable $y$ is called the dependent variable, because it depends on $x$. Once you have chosen a value for $x$, the value of $y$ is determined.

**distribution** For a numerical or categorical data set, the distribution tells you how many of each value or each category there are in the data set.

**domain** The domain of a function is the set of all of its possible input values.

**E**

**elimination** A method of solving a system of two equations in two variables where you add or subtract a multiple of one equation to another in order to get an equation with only one of the variables (thus eliminating the other variable).

**equivalent equations** Equations that have the exact same solutions are equivalent equations.

**equivalent systems** Two systems are equivalent if they share the exact same solution set.

**exponential function** An exponential function is a function that has a constant growth factor. Another way to say this is that it grows by equal factors over equal intervals. For example, $f(x) = 2 \cdot 3^x$ defines an  exponential function. Any time $x$ increases by 1, $f(x)$ increases by a factor of 3.

**F**

**factored form (of a quadratic expression)** A quadratic expression that is written as the product of a constant times two linear factors is said to be in factored form. For example, $2(x - 1)(x + 3)$ and $(5x + 2)(3x - 1)$ are both in factored form.

**five-number summary** The five-number summary of a data set consists of the minimum, the three quartiles, and the maximum. It is often indicated by a box plot like the one shown, where the minimum is 2, the three quartiles are 4, 4.5, and 6.5, and the maximum is 9.

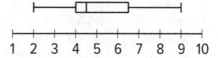

**function** A function takes inputs from one set and assigns them to outputs from another set, assigning exactly one output to each input.

**function notation** Function notation is a way of writing the outputs of a function that you have given a name to. If the function is named $f$ and $x$ is an input, then $f(x)$ denotes the corresponding output.

**G**

**graph of a function** The graph of a function is the set of all of its input-output pairs in the coordinate plane.

**growth factor** In an exponential function, the output is  multiplied by the same factor every time the input increases by one. The multiplier is called the growth factor.

**growth rate** In an exponential function, the growth rate is the fraction or percentage of the output that gets added every time the input is increased by one. If the growth rate is 20% or 0.2, then the growth factor is 1.2.

---
## H

**horizontal intercept** The horizontal intercept of a graph is the point where the graph crosses the horizontal axis. If the axis is labeled with the variable $x$, the horizontal intercept is also called the $x$-intercept. The horizontal intercept of the graph of $2x + 4y = 12$ is (6,0).

The term is sometimes used to refer only to the $x$-coordinate of the point where the graph crosses the horizontal axis.

---
## I

**increasing (function)** A function is increasing if its outputs get larger as the inputs get larger, resulting in an upward sloping graph as you move from left to right.

A function can also be increasing just for a restricted range of inputs. For example the function $f$ given by $f(x) = 3 - x^2$, whose graph is shown, is increasing for $x \leq 0$ because the graph slopes upward to the left of the vertical axis.

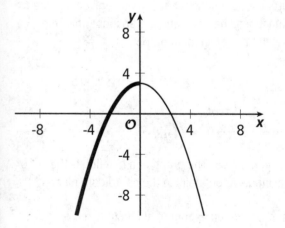

**independent variable** A variable representing the input of a function.

The equation $y = 6 - x$ defines $y$ as a function of $x$. The variable $x$ is the independent variable, because you can choose any value for it. The variable $y$ is called the dependent variable, because it depends on $x$. Once you have chosen a value for $x$, the value of $y$ is determined.

**intercept** A point on the graph of a function which is also on one of the axes.

**inverse (function)** Two functions are inverses to each other if their input-output pairs are reversed, so that if one functions takes $a$ as input and gives $b$ as an output, then the other function takes $b$ as an input and gives $a$ as an output. You can sometimes find an inverse function by reversing the processes that define the first function in order to define the second function.

**irrational number** An irrational number is a number that is not rational. That is, it cannot be expressed as a positive or negative fraction, or zero.

---
## L

**linear function** A linear function is a function that has a constant rate of change. Another way to say this is that it grows by equal differences over equal intervals. For example, $f(x) = 4x - 3$ defines a linear function. Any time $x$ increases by 1, $f(x)$ increases by 4.

**linear term** The linear term in a quadratic expression (In standard form) $ax^2 + bx + c$, where $a$, $b$, and $c$ are constants, is the term $bx$. (If the expression is not in standard form, it may need to be rewritten in standard form first.)

---
## M

**maximum** A value of a function that is greater than or equal to all the other values, corresponding to the highest point on the graph of the function.

**minimum** A value of a function that is less than or equal to all the other values, corresponding to the lowest point on the graph of the function.

**model** A mathematical or statistical representation of a problem from science, technology, engineering, work, or everyday life, used to solve problems and make decisions.

## N

**negative relationship** A relationship between two numerical variables is negative if an increase in the data for one variable tends to be paired with a decrease in the data for the other variable.

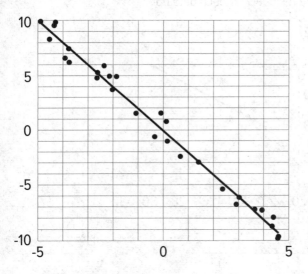

**non-statistical question** A non-statistical question is a question which can be answered by a specific measurement or procedure where no variability is anticipated, for example:

- How high is that building?
- If I run at 2 meters per second, how long will it take me to run 100 meters?

**numerical data** Numerical data, also called measurement or quantitative data, are data where the values are numbers, measurements, or quantities. For example, the weights of 10 different dogs are numerical data.

## O

**outlier** A data value that is unusual in that it differs quite a bit from the other values in the data set. In the box plot shown, the minimum, 0, and the maximum, 44, are both outliers.

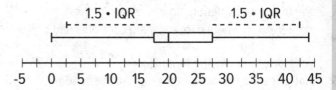

## P

**perfect square** A perfect square is an expression that is something times itself. Usually we are interested in situations where the something is a rational number or an expression with rational coefficients.

**piecewise function** A piecewise function is a function defined using different expressions for different intervals in its domain.

**positive relationship** A relationship between two numerical variables is positive if an increase in the data for one variable tends to be paired with an increase in the data for the other variable.

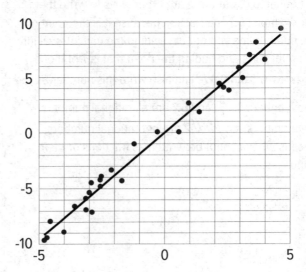

## Q

**quadratic equation** An equation that is equivalent to one of the form $ax^2 + bx + c = 0$, where $a$, $b$, and $c$ are constants and $a \neq 0$.

**quadratic expression** A quadratic expression in $x$ is one that is equivalent to an expression of the form $ax^2 + bx + c$, where $a$, $b$, and $c$ are constants and $a \neq 0$.

**quadratic formula** The formula $x = \dfrac{-b \pm \sqrt{b^2 - 4ac}}{2a}$ that gives the solutions of the quadratic equation $ax^2 + bx + c = 0$, where $a$ is not 0.

**quadratic function** A function where the output is given by a quadratic expression in the input.

## R

**range** The range of a function is the set of all of its possible output values.

**rational number** A rational number is a fraction or the opposite of a fraction. Remember that a fraction is a point on the number line that you get by dividing the unit interval into $b$ equal parts and finding the point that is $a$ of them from 0. We can always write a fraction in the form $\dfrac{a}{b}$ where $a$ and $b$ are whole numbers, with $b$ not equal to 0, but there are other ways to write them. For example, 0.7 is a fraction because it is the point on the number line you get by dividing the unit interval into 10 equal parts and finding the point that is 7 of those parts away from 0. We can also write this number as $\dfrac{7}{10}$.

The numbers 3, $-\dfrac{3}{4}$, and 6.7 are all rational numbers. The numbers $\pi$ and $-\sqrt{2}$ are not rational numbers, because they cannot be written as fractions or their opposites.

**relative frequency table** A version of a two-way table in which the value in each cell is divided by the total number of responses in the entire table or by the total number of responses in a row or a column. The table illustrates the first type for the relationship between the condition of a textbook and its price for 120 of the books at a college bookstore.

| | $10 or Less | More than $10 but Less than $30 | $30 or More |
|---|---|---|---|
| new | 0.025 | 0.075 | 0.225 |
| used | 0.275 | 0.300 | 0.100 |

**residual** The difference between the $y$-value for a point in a scatter plot and the value predicted by a linear model. The lengths of the dashed lines in the figure are the residuals for each data point.

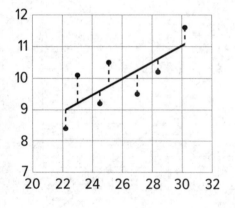

## S

**skewed distribution** A distribution where one side of the distribution has more values farther from the bulk of the data than the other side, so that the mean is not equal to the median. In the dot plot shown, the data values on the left, such as 1, 2, and 3, are further from the bulk of the data than the data values on the right.

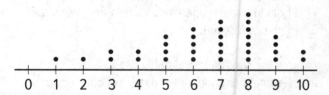

**solution to a system of equations** A coordinate pair that makes both equations in the system true.

On the graph shown of the equations in a system, the solution is the point where the graphs intersect.

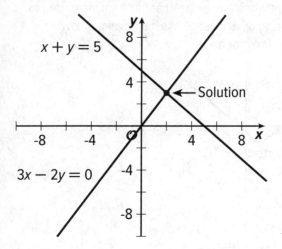

$$x + y = 5$$

$$3x - 2y = 0$$

**solutions to a system of inequalities** All pairs of values that make the inequalities in a system true are solutions to the system. The solutions to a system of inequalities can be represented by the points in the region where the graphs of the two inequalities overlap.

**standard deviation** A measure of the variability, or spread, of a distribution, calculated by a method similar to the method for calculating the MAD (mean absolute deviation). The exact method is studied in more advanced courses.

**standard form (of a quadratic expression)** The standard form of a quadratic expression in $x$ is $ax^2 + bx + c$, where $a$, $b$, and $c$ are constants, and $a$ is not 0.

**statistic** A quantity that is calculated from sample data, such as mean, median, or MAD (mean absolute deviation).

**statistical question** A statistical question is a question that can only be answered by using data and where we expect the data to have variability, for example:

- Who is the most popular musical artist at your school?
- When do students in your class typically eat dinner?
- Which classroom in your school has the most books?

**strong relationship** A relationship between two numerical variables is strong if the data is tightly clustered around the best fit line.

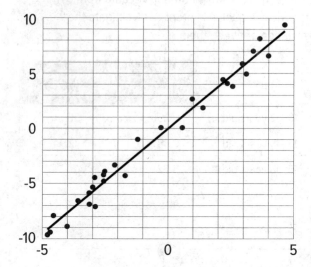

**substitution** Substitution is replacing a variable with an expression it is equal to.

**symmetric distribution** A distribution with a vertical line of symmetry in the center of the graphical representation, so that the mean is equal to the median. In the dot plot shown, the distribution is symmetric about the data value 5.

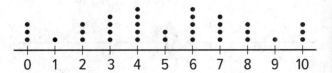

**system of equations** Two or more equations that represent the constraints in the same situation form a system of equations.

**system of inequalities** Two or more inequalities that represent the constraints in the same situation form a system of inequalities.

## T

**two-way table** A way of organizing data from two categorical variables in order to investigate the association between them.

| | has a cell phone | does not have a cell phone |
|---|---|---|
| 10–12 years old | 25 | 35 |
| 13–15 years old | 38 | 12 |
| 16–18 years old | 52 | 8 |

## U

**uniform distribution** A distribution which has the data values evenly distributed throughout the range of the data.

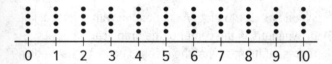

## V

**variable (statistics)** A characteristic of individuals in a population that can take on different values

**vertex (of a graph)** The vertex of the graph of a quadratic function or of an absolute value function is the point where the graph changes from increasing to decreasing or vice versa. It is the highest or lowest point on the graph.

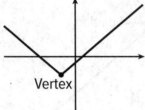

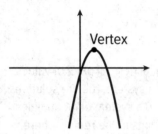

**vertex form (of a quadratic expression)** The vertex form of a quadratic expression in $x$ is $a(x - h)^2 + k$, where $a$, $h$, and $k$ are constants, and $a$ is not 0.

**vertical intercept** The vertical intercept of a graph is the point where the graph crosses the vertical axis. If the axis is labeled with the variable $y$, the vertical intercept is also called the $y$-intercept.

Also, the term is sometimes used to mean just the $y$-coordinate of the point where the graph crosses the vertical axis. The vertical intercept of the graph of $y = 3x - 5$ is (0, -5), or just -5.

## W

**weak relationship** A relationship between two numerical variables is weak if the data is loosely spread around the best fit line.

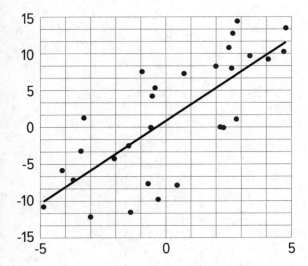

## Z

**zero (of a function)** A zero of a function is an input that yields an output of zero. If other words, if $f(a) = 0$ then $a$ is a zero of $f$.

**zero product property** The zero product property says that if the product of two numbers is 0, then one of the numbers must be 0.

# Index